CRATER TO THE CREGGAN

COLONEL-IN-CHIEF

Her Majesty Queen Elizabeth The Queen Mother

DEPUTY COLONELS-IN-CHIEF

HRH The Princess Margaret, Countess of Snowdon
HRH The Princess Alice, Duchess of Gloucester

Her Majesty Queen Elizabeth The Queen Mother

Colonel – in – Chief

A portrait by Richard Stone

CRATER TO THE CREGGAN

A History of

THE ROYAL ANGLIAN REGIMENT 1964–1974

BY

MICHAEL BARTHORP

LEO COOPER · LONDON

First published in Great Britain 1976 by
LEO COOPER LTD

196 Shaftesbury Avenue, London WC2H 8JL

ISBN 0 85052 212 9

Printed in Great Britain by
Ebenezer Baylis & Son Limited
The Trinity Press, Worcester, and London

CONTENTS

MAPS

ILLUSTRATIONS

ACKNOWLEDGEMENTS

The sources used for this history have been regimental documents and minutes, battalion diaries and accounts written by officers, official citations, newspaper reports, letters, and the Regimental magazine, *Castle*. For background material the following published works have been of value: Gregory Blaxland's *The Regiments Depart* (1971), Julian Paget's *Last Post: Aden 1964–67* (1969), J. C. M. Baynes' *The Soldier in Modern Society* (1972), *The Sunday Times'* 'Ulster' (1972); the *Journal of the Royal United Services Institute*; and for historical matters, H. M. Chichester and G. Burges Short's *Records and Badges of the British Army* (1900) and the *Journal of the Society for Army Historical Research*. Where quotations have been used, their sources are noted either in the text or a footnote.

In Chapter 9 the interests of security of members of the Regiment have required the omission of certain names, facts and incidents. To some degree this has prevented full credit being given to some of the notable services performed by officers and men.

I acknowledge with gratitude information and help given by the following officers in reply to my enquiries: Lieutenant-Colonel K. Burch MBE, Lieutenant-Colonel T. D. Dean, Brigadier R. E. J. Gerrard-Wright OBE, Lieutenant-Colonel T. Holloway, Major-General M. W. Holme CBE, MC, Major J. K. Parker, Captain P. P. Rawlins, Lieutenant-Colonel M. E. Thorne OBE, and Lieutenant-Colonel P. Worthy. I owe a particular debt to Brigadier J. B. Akehurst, at whose suggestion I undertook the work, for information, ideas and encouragement, and to Lieutenant-Colonel C. R. Murray Brown DSO who, as the link between Regiment and author, has replied to many importunate letters and has produced most of the Appendices.

I must also acknowledge the pioneering work done by Lieutenant-Colonel P. G. Upcher DSO towards the production of a history of the Regiment, and the many comments and suggestions submitted by members of the Regiment as a result both of his work and my first draft.

Perhaps my greatest debt of all is due to the officers and soldiers, many of them anonymous, without whose contributions to the *Castle* this story could not have come to life.

Finally, it could never have been completed without the patient encouragement of my wife, who has watched its production with a stoicism worthy of a soldier's wife and daughter.

M.J.B.

FOREWORD

THE PURPOSE OF this short history covering the first ten years in the life of the Royal Anglian Regiment is twofold. Firstly, to describe the various stages of evolution of the Regiment and secondly, to record the main events and achievements during the period.

In this way, I hope that everyone connected with the Regiment, either in the previous Regiments or the civilian community in our large Regimental area, will be able to understand more clearly the organization and motivation of their modern County Regiment within the Queen's Division of Infantry.

The story begins at the time of the withdrawal from Aden, our last major Colonial territory, in which three of our four battalions played a significant part, and ends with the Northern Ireland emergency which has involved all our regular battalions to a greater or lesser degree and still continues after six years. The success of our battalions in this particularly difficult type of security operation will long be remembered. At least as significant is the part played by senior officers of the Regiment, who so far have held more key command appointments in Northern Ireland, Aden, the Persian Gulf and in Muscat and Oman than any other regiment.

Interspersed with operations in Aden and Northern Ireland the 1st, 2nd and 3rd Battalions have all undertaken the rôle in BAOR of the mechanized infantry battalion, which is the most complex fighting unit in the modern Army, containing a wider range and number of weapons, vehicles and communications than any unit of another arm. This rôle places great responsibility on junior leaders which has proved of considerable benefit for the exacting task in Northern Ireland where good junior leadership is so important.

Since formation, the most important change in the Regiments' organization has been the reduction in the regular operational strength from four battalions to three and the formation of three volunteer battalions. Although our 4th Battalion was reduced to company strength in 1971 and was finally disbanded in July, 1975, the Tigers live on in spirit and men from Leicestershire serve in all battalions, while the County continues to give the Regiment magnificent support.

When our Colonel-in-Chief presented identical Colours to the Regular Battalions and the 5th (Volunteer) Battalion on 12 July, 1974, in the tenth year of the Regiment's existence, it was brought home to everyone present that we are one Regiment and a worthy successor to all the previous Regiments whose traditions and standards we have inherited. All ranks have worked hard to achieve these high standards of which we can be justly proud. We look forward with confidence to the Regiment continuing to make its mark in the future wherever it is needed in this uncertain world, both in the military and sporting spheres, and to a continuing support from all the Counties in our large Regimental area.

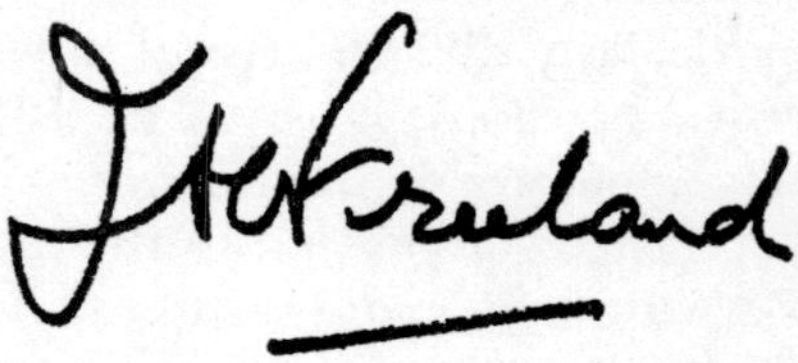

COLONEL OF THE REGIMENT

August, 1975

CRATER TO THE CREGGAN

The Family Tre

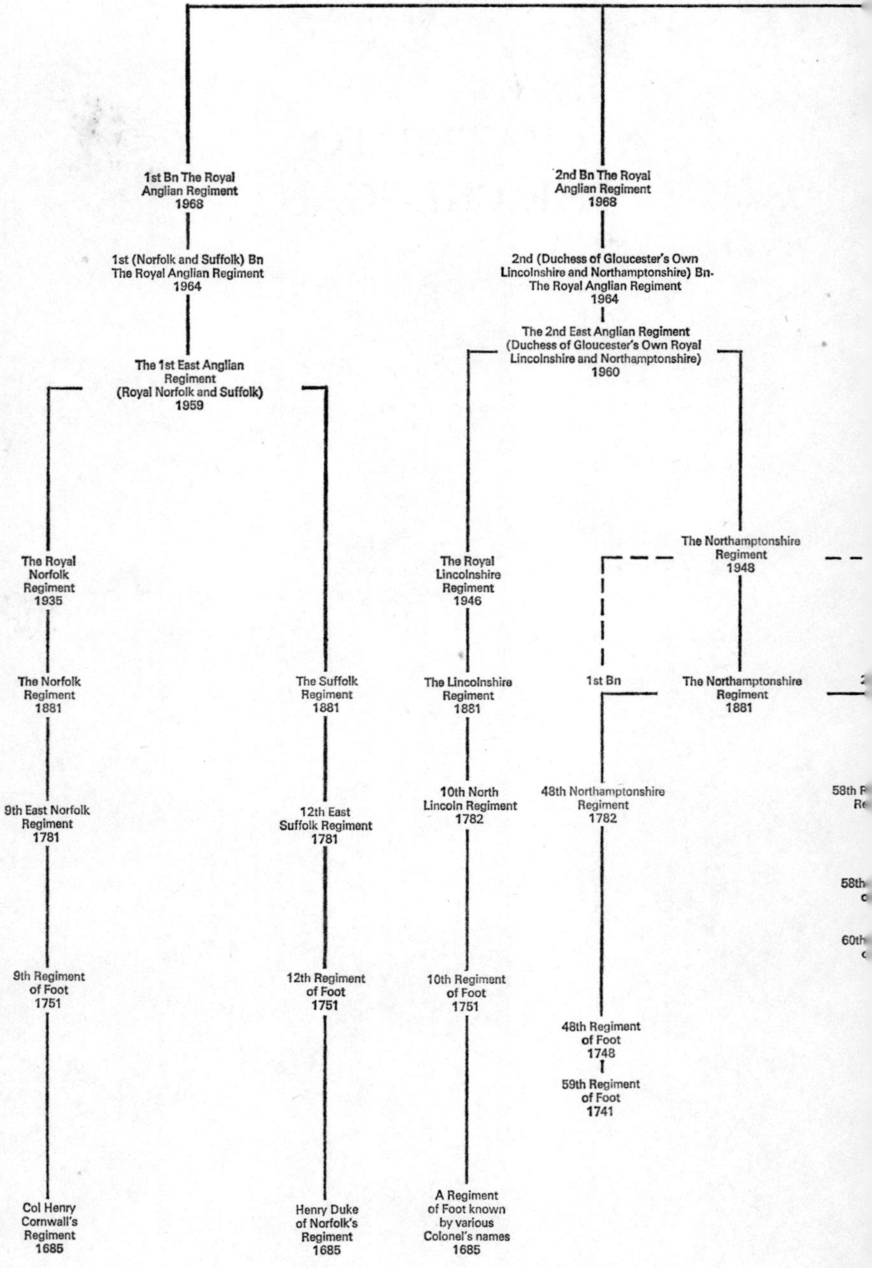

ANGLIAN REGIMENT

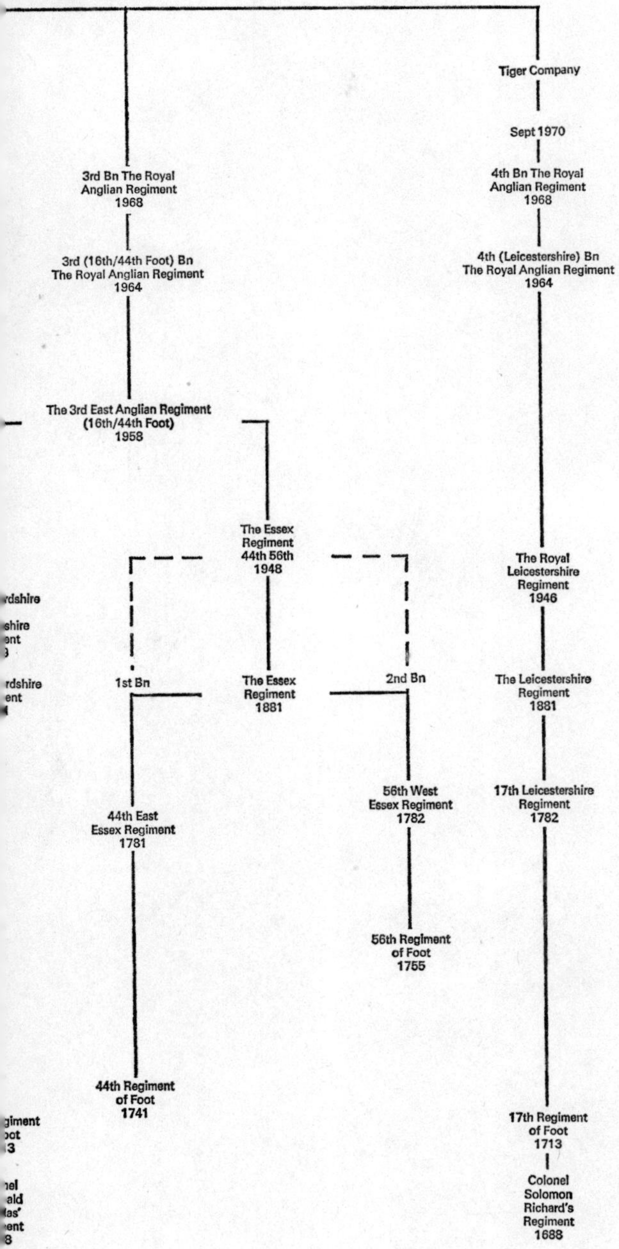

CHAPTER 1

CO-OPERATIVE EVOLUTION

> 'The British Infantry are the best in the world.
> Fortunately there are not many of them.'
>
> Marshal Bugeaud

ON 1 SEPTEMBER, 1964, men of four regiments of the Infantry of the Line marched on to parade grounds in England, Germany and the Arabian Peninsula. In the ranks stood townsmen and countrymen, some Cockneys, a Scot or two, perhaps some Welsh, inevitably a few Irish, which scarcely a Regiment of the Line has ever been without, but mostly they were men from the eastern counties of England; from Lincolnshire and Leicestershire in the north, to Essex and Hertfordshire in the south; from Bedfordshire and Northamptonshire in the west, to Norfolk and Suffolk in the east, not forgetting Rutlandshire, Huntingdonshire and Cambridgeshire in the centre. In khaki and white, in scarlet and blue, men from eleven counties paraded. Bands played, commands rang out, exhortations were made, and then they marched off; not, as they had marched on, as members of four separate entities, but now as part of one corporate whole. In the interim a metamorphosis had occurred. The parades had marked the entry of a new name to the Army List and a new regiment had joined the Infantry of the Line—the Royal Anglian.

It probably did not cross the minds of many as they marched off parade that they had just achieved a 'first' in the three hundred years history of the British Army. But this they had done. They had been the first to participate in the birth of a new concept for the organization of the Infantry, the formation of what was then known as a 'Large Regiment'. Since this concept now appears to be a dead issue, no Large Regiment having been formed since 1968, it is of historical interest to trace the events and thinking that led up to its introduction, the discussions that ensued within the regiments affected and the measures taken to effect it.

The years following the Second World War found the County Regiments, which hitherto had had two Regular battalions, not only reduced to one Regular and one or two Territorial battalions, but grouped for administrative purposes into territorial brigades. That which concerns this story was termed the East Anglian Brigade and contained five regiments: those of Norfolk, Suffolk, Bedfordshire and Hertfordshire, Essex, and Northamptonshire. The Government decision in 1957 to end conscription and revert to an all-Regular Army, though welcomed by Regulars and potential conscripts alike, brought a sting in its tail: a reduction in the strength of the Army from 400,000 to 180,000.

In the Infantry the method to be adopted was the amalgamation of pairs of regiments, rather than disbandment. Because of their lack of recruiting potential, all regiments of the East Anglian Brigade were singled out for this dubious honour; Royal Norfolk with Suffolk, Essex with Bedfordshire and Hertfordshire, and, to provide a partner for the Northamptons, the Royal Lincolns were enticed from the Midland Brigade. Between 1958–1960 these marriages were celebrated; not perhaps such shot-gun affairs as occurred elsewhere in the Infantry, but nevertheless painful and requiring much tolerance and good will. By the end of 1960, then, the East Anglian Brigade was reduced to three regiments, all separate entities, each with its own Regular or 1st Battalion, but all bearing the same name, though with certain appendages tacked on behind to placate those fearful of the impact of lost traditions, as follows:

1st East Anglian Regiment (Royal Norfolk and Suffolk)
2nd East Anglian Regiment (Duchess of Gloucester's Own Royal Lincolnshire and Northamptonshire)
3rd East Anglian Regiment (16th/44th Foot).

It could perhaps be said that the only true East Anglians of this arrangement adopted a straightforward addendum; the most far-flung elements showed their distaste by choosing an appendage of inordinate length; while the junior regiment, either from a wish for brevity or in a display of extreme nostalgia, reverted to the style of earlier years!

Under this somewhat bizarre nomenclature, the new partners went forth to life together, hoping for a period in which to achieve, in the words of the marriage service, 'the mutual society, help and comfort, that the one ought to have for the other, both in prosperity and adversity'. However, in the corridors of power, minds were soon again at work, seeking to create greater flexibility to meet the changing needs of strategy. This meant in effect that, notwithstanding the very considerable upheavals caused by the amalgamations, some further reorganization of the Infantry was necessary so that the number of Regular battalions, then 60, might be varied from time to time with the minimum disturbance; the variation envisaged ranged from a contraction to 50 battalions to an expansion of 70. The nub of this requirement was the term 'minimum disturbance'. But how was this to be achieved without the finality of disbandment, the limbo of suspended animation, or further amalgamations, not only of regiments which had avoided matrimony in 1958–60, but possibly also of further unions for those who had already been hauled to the altar of economy?

The Army Council decided that disturbance could best be avoided by first inflicting another major, but this time voluntary, convulsion on the somewhat battered Regiments of the Line, after which reductions could then follow less painfully; a form of being cruel to be kind. The convulsion chosen was the conversion of the territorial brigades into large regiments, which at no time should contain less than three battalions each. In 1962 a committee under the chairmanship of Lieutenant-General Sir Roger Bower was set up to advise on how this step should be implemented.

The concept of these large regiments was that each should have one head, its Colonel, one name, with its battalions numbered consecutively, one body, with all ranks on a common roll, and one home, its depot. The aim was to encourage an organization which could perpetuate the essential features of the regimental system, absorb its traditions, territorial links and customs, and withstand the stresses of change in the size and shape of the Army as a whole.

The viability, to use a popular word of the time, of the concept was the organization of the regiment into that number of battalions necessary to ensure that, if and when cuts in the number of battalions were applied, the large regiment would still be capable of continuing to function efficiently as such.

It became clear to the Bower Committee that, if this viability was to be achieved and to ensure that all ranks could serve at least one tour in three in the United Kingdom, the optimum size for a large regiment was four battalions. As seven of the territorial brigades of Infantry, of which the East Anglian was one, contained only three regiments, some regrouping of these brigades was necessary. The Committee decided that, for the fourth regiment in the East Anglian Brigade, geographically, and therefore for recruiting reasons, the Royal Leicestershire Regiment (from the Forester Brigade) was the most appropriate.

The views of Leicestershire as to its affinity with East Anglia seem not to have been sought.

The Committee agreed that the time (1962) was too soon after the earlier amalgamations to press at once for the conversion to large regiments, and that anyway further cuts in the Infantry were unlikely in the near future. They therefore felt that large regiments should be allowed to develop, as far as possible, by 'co-operative evolution', but that, in case cuts were demanded by Government at short notice, the Infantry must move some way towards the large regiment in order to prevent the total extinction of any regiment.

On 31 May, 1962, a copy of the Bower Report was sent to the Representative Colonel of the East Anglian Brigade, Lieutenant-General Sir Reginald Denning, Colonel of the 3rd East Anglian Regiment. On 14 November the Army Council's plan for the Infantry, based on the Bower Report, was promulgated to all concerned, followed, on 28 February, 1963, by a further letter, explaining the concept in detail, stressing its voluntary nature, and also stating that 'the adoption of the Large Regiment concept will in no way make the Regiments concerned more liable in the event of making reductions'.

For the three regiments of the East Anglian Brigade, all of them amalgamations and all with a common title, compliance with the Army Council's wishes posed fewer anxieties than for some regiments in other brigades, who seized upon the voluntary nature of the affair and dug in within their county boundaries. The chief problem was the Royal Leicesters. Here was a well-recruited regiment, proudly known as the 'Tigers', hitherto unscarred by the axe of flexibility, facing, by virtue of its junior position in its new brigade, the possibility of extinction through no fault of its own. To many of its members its identity must suddenly have seemed to be precarious. To its great credit, and aided with goodwill and understanding of its position by the East Anglian Council of Colonels, the Regiment took a constructive view and its then Colonel, Lieutenant-General Sir Colin Callander, wrote, 'I do not think we can ever produce a true regimental spirit in our Brigade until we form a large Regiment of four battalions'. He declined, however, with good reason, a suggestion that, until a large regiment was formed, his own might be transmuted into the 4th East Anglian Regiment. So, on 1 May, 1963, the Royal Leicestershire Regiment joined the East Anglian Brigade, under its own identity, but adopting the Brigade cap badge and other items of dress common within the Brigade. The scene was now set for the next act.

After a meeting later in May under the chairmanship of Colonel Denny, the Brigade Colonel, to whom so much of the successful outcome of the problem was due, the Commanding Officers of the four Regular battalions recommended to the Council of Colonels that the Brigade should convert to a large regiment as soon as possible, subject to certain conditions about the maintenance of county connections. The Council of Colonels agreed, but when General Denning reported their decision to the then Chief of the Imperial General Staff, General Sir Richard Hull, he also emphasized that his Council's decision was dependent on other Brigades also converting. He was told that the Army Council was firmly determined that all Brigades should do so. In the event, however, only six out of thirteen did so, and three of these took the step under somewhat different circumstances, as will be seen later.

The title of the new regiment naturally provoked much discussion. A suggestion that it might be linked to a royal personage* received some support but there was more enthusiasm for a geographical appellation. The Royal Regiment of Anglia found favour among the Commanding Officers. East Anglian was territorially imprecise but by omitting the 'East' the local sensibilities of the middle Angles were mollified, particularly if the 'Royal' title, already owned by three of the old county regiments, could be prefixed. So Royal Anglian it was to be; a not unpleasing, seemingly easily memorized name, with, in addition, little opportunity for solecisms, as might have occurred with an abbreviated version of Royal East Anglian Regiment. Even so, it was not all plain sailing. For some reason best known to themselves, some sections of the Press elected to bestow

* Queen Elizabeth's Own Royal Regiment was suggested.

upon the Regiment an ecclesiastical air by referring to it as 'Royal Anglican'.

The proviso about maintaining county connections mentioned earlier included the use of subsidiary titles in parentheses after each battalion's number. These were agreed, although the CIGS was unable to guarantee their use in perpetuity as had been desired by the Council of Colonels. The four battalions thus started life as:

1st (Norfolk and Suffolk) Battalion
2nd (Duchess of Gloucester's Own Lincolnshire and Northamptonshire) Battalion
3rd (16th/44th Foot) Battalion
4th (Leicestershire) Battalion.

Another symptom of the transitional mood of the change-over was the unanimous wish of all battalions to retain the collar badges of their former regiments; this was in conflict with the concept of the large regiment but was perhaps understandable at the time. They represented legacies of the past, as indeed did the other regimental insignia, which will be discussed in the next chapter.

A rather more delicate matter concerned the three Royal ladies who had been Colonels-in-Chief of the three East Anglian Regiments: HM Queen Elizabeth the Queen Mother of the 3rd, who previously had been Colonel-in-Chief of the Bedfords since 1949; HRH The Princess Margaret, Countess of Snowdon, of the 1st, and of the Suffolk Regiment since 1953; and HRH The Duchess of Gloucester of the 2nd, who had been Colonel-in-Chief of the Northamptons since 1937. Clearly the new regiment could have only one Colonel-in-Chief, but equally the former regiments were anxious not to lose the connections with their own Colonels-in-Chief, who had always taken such interest in their well-being. After much discussion, the Army Council submitted to the Palace the request that HM Queen Elizabeth the Queen Mother should be appointed Colonel-in-Chief of the Royal Anglian Regiment, and that HRH The Princess Margaret and HRH

1st (Norfolk and Suffolk) Bn

2nd (Duchess of Gloucester's Own Lincolnshire and Northamptonshire) Bn

3rd (16th/44th Foot) Bn

4th (Leicestershire) Bn

Collar badges of the three East Anglian Regiments and the Royal Leicestershire Regiment adopted in 1964 for use by the Royal Anglian Regiment on formation.

The Duchess of Gloucester be appointed Deputy Colonels-in-Chief. To this proposal HM The Queen most graciously gave her consent, and on 26 November, 1964, at a reception held at St James's Palace to cement the formation of the new regiment, some 1,000 officers and their wives were honoured by the presence of the Royal ladies attending a regimental function for the first time in their new appointments.

The first Colonel of the Regiment was General Denning, who had done so much to set the new regiment on the right course, and to assist him in his duties, there were appointed Deputy Colonels to look after the interests of each battalion:

Lieutenant-General Sir Richard Goodwin, Brigadier Oulton, Brigadier Paton and Major-General Sir Douglas Kendrew.

Regimental Headquarters and the Depot were located at Bury St Edmunds, with outstations to watch over the concerns of the seven former county regiments at Norwich, Bury St Edmunds, Lincoln, Northampton, Bedford, Warley and Leicester.

No mention has been made here of the effect of the reorganization on the Territorial battalions of the former regiments, which in fact at this stage was negligible, but their part will be dealt with in a later chapter.

And so the preliminaries—the plans, the meetings, the discussions—moved gradually to the end product: Formation Day, 1 September, 1964. Other regiments of other brigades may have viewed with amusement, astonishment, or even scorn, the East Anglian Brigade's alacrity to volunteer for what may have seemed to them an unnecessary loss of treasured identity. The Royal Anglians' wisdom in going to the large regiment, and their subsequent success with the concept, can best be judged at the end of this story. In the interim, however, it might be noted that they were not alone and were indeed in good company. The Green Jackets Brigade, who were never short of influence in high places to plead their cause when confronted with disagreeable measures, were moving in the same direction and converted to a large regiment, the Royal Green Jackets, on 1 January, 1966. They had always prided themselves on being forward-looking, whereas perhaps a fault of the red-coated Infantry was a tendency to look backwards. The Royal Anglians, at least, could not be accused of that.

CHAPTER 2

HEROES OF ANTIQUITY

'The men that fought at Minden, they was rookies in
their time—
So was them that fought at Waterloo!
All the 'ole command, yuss, from Minden to Maiwand,
They was once dam' sweeps like you!'

Rudyard Kipling, *The Men that fought at Minden.*

As THE BATTALIONS of the new regiment marched confidently off to their future, so did their ghostly forbears—the shades of nine, then seven, and finally four proud regiments—fade away, if not into oblivion, at least out of the consciousness of the modern Royal Anglians. And rightly so; for any innovation, modernity and the future is of the essence if it is to succeed; to look back is to defeat its purpose.

Yet the Royal Anglian was not a new regiment, not in the sense that it was raised from nothing, as its forbears had once been. The very fact that it had forbears precluded it from being new. Its formation was simply a development, one of several, in the continuing and unbroken history of the British Infantry. So before following the fortunes of the Royal Anglians it seems fitting to look for a moment at those forbears, if only to ensure that, in the future, their long and noble record is not entirely forgotten, and that soldiers yet to serve are aware whence came the insignia they wear and the traditions they find themselves perpetuating.

The story starts in 1685, the year of the rebellion of the Duke of Monmouth against King James II, when nine regiments of infantry were added to the English establishment. Of these, three subsequently took precedence in the Line as the 9th, 10th and 12th Regiments of Foot. Three years later further regiments were raised, amongst them two which were later numbered 16th and 17th Foot. The years passed, the years of Marlborough's wars, and on into the eighteenth century until 1741, when the establishment was again increased, thus bringing to birth the 44th and 48th Regiments of Foot. The increasing power of France demanded a still larger British Army, and so, in 1755, came the 56th and 58th Regiments, to complete the muster of the Royal Anglian's forefathers.

In 1782 a letter was sent by the Deputy Adjutant General to all Colonels of Regiments of Foot, asking if they had 'any particular connexion [*sic*] or attachment to any particular county, or any reason for bearing the name of any particular county' in addition to their number. Generally the scheme evoked little enthusiasm from regiments but nevertheless it went forward, though the county titles subsequently allotted were not always those selected by regiments. The 9th and 12th Foot both chose Norfolk, but the former had to be content with East Norfolk and the latter with East Suffolk. The 10th asked for Northumberland but was granted North Lincoln. The 58th suggested Warwickshire, Lancashire, or, ignoring the point of the letter entirely, the title of 'The Black Regiment', from the colour of their facings; they were sternly told that their portion would be Rutlandshire. The Colonel of the 44th, Major-General Rainsford, had connections with Essex and these were perpetuated, though he had to be content with the eastern half of the county, the western going to the 56th, whose own choice is not recorded. The wishes of the 16th, 17th and 48th are not on evidence—perhaps like many they were content with their numbers—but they were, in the event affiliated, the 16th to Buckinghamshire, changed in 1809 by agreement with the 14th Foot to Bedfordshire, the 17th to Leicestershire, and the 48th to Northamptonshire.*

* 'County Names for Regiments 1782', *Journal of the Society for Army Historical Research*, Volume XXXVI, pp. 34–8.

Despite all this, the regiments pursued their way for nearly another hundred years under their time-honoured numbers, the county titles having little or no significance. There was no regimental presence in the counties, depots were situated where most convenient, and recruits came from wherever they could be gleaned. Then in 1881 came a reform as disturbing and unpopular as any of the changes perpetrated upon the Infantry in recent years. The treasured numbers were abolished, and, in the case of English Infantry, all regiments were to be firmly linked to and styled under the names of the counties of England, each with two Regular battalions and one or more Militia and Volunteer battalions. For the older regiments—the 1st to the 25th Foot—the step was not so terrible, since each already had two Regular battalions. For the others, which included the 44th, 48th, 56th and 58th, it meant the loss of precious individuality and union with a stranger, a forerunner in fact of the amalgamations that began in 1958.

So the county regiments evolved. The 9th Foot was transformed into The Norfolk Regiment, granted the Royal title in 1935; the 10th into The Lincolnshire Regiment, subsequently becoming Royal after the Second World War; the 12th Foot, The Suffolk Regiment, the 16th, The Bedfordshire Regiment, a title later expanded to include Hertfordshire; the 17th, The Leicestershire Regiment, also made Royal after the Second World War; the 44th and 56th became the 1st and 2nd Battalions of The Essex Regiment, and the 48th and 58th, the 1st and 2nd Battalions of The Northamptonshire Regiment. In the latter cases individual regiments were transformed overnight into battalions of a larger regiment, exactly as occurred in 1964, though, it must be said, with considerably more distress and concern for loss of the past than was evident at the birth of the Royal Anglian Regiment.

The demands of two World Wars so fortified the links between regiments and counties that, when the Infantry was next convulsed by the reductions of 1958–60, the prospective loss of county titles caused almost as much heart-rending as had the abolition of the numbers in 1881. The administrative convenience of grouping county regiments into territorial brigades, introduced in 1947, in no way sweetened the pill of amalgamation. Regimental identity was the thing, and living memories forgot that many identities had already been lost in 1881.

Nevertheless progress must be served, so the county titles were marched off into parentheses and three East Anglian Regiments took the stage with one at least, Lincolns and Northamptons, blinking somewhat uncertainly under the light of a cognomen to which it had no conceivable geographical claim. Geography, however, was of little account in the deliberations of Whitehall, as was seen when the Tigers were hunted out of their Midland forests into the fens and fields of East Anglia, whose traditional boundaries grew 'wider still and wider'. So the old regiments were manoeuvred into position to await, with some trepidation, the next onslaught on a system treasured and envied by so many. When it came in the form of a polite 'wish', it seemed, after so much upheaval, logical and sensible enough to be complied with. Thus, over 279 years, nine regiments became one.

The determination of the Royal Anglian Regiment to be a forward-looking entity did not preclude the preservation of time-honoured insignia, traditions and customs of its forbears. The Castle and Key of Gibraltar, granted to the 12th, 56th and 58th Regiments for their part in the Great Siege of 1779–1783, forms the central motif of the cap and collar badges, placed above a regimental scroll and set upon the eight-pointed star borne on the old badges of the Royal Lincolns and Bedfords. The Royal Tiger of Hindoostan, awarded to the 17th 'as a lasting testimony to the exemplary conduct of the corps during its service in India from 1804 to 1823', is engraved upon the buttons in brass within the unbroken laurel wreath, commemorating a January morning in 1777 when 250 of the 17th broke their way out of a ring of 3,000 American revolutionaries under George Washington at Princetown. This is the second button design adopted by the Regiment, the first having the Star and Castle.

Lanyards of yellow for the 1st Battalion recall the facings of that colour worn by the 9th and 12th Foot; of black for the 2nd Battalion, those of the 58th; for the 3rd Battalion of purple, of the shade once known as Pompadour from the favourite colour of the best-loved mistress of Louis XV, the old facings of the 56th; while the 4th Battalion retain the red, pearl grey and black braided lanyard of the Royal Leicesters.

Upon the Regimental Colour of each battalion are emblazoned ancient emblems: the figure of Britannia, granted to the 9th Foot by Queen Anne for its gallantry at Almanza in Spain in 1707; the Sphinx with Egypt, granted to the 58th, who held the right of the line at Alexandria in 1801, and to the 10th, who came from India in the same year to march across the desert to Cairo; the Eagle of the French 62nd Regiment taken at Salamanca in 1812 by the 44th, who were also present in Egypt in 1801 and were granted the Sphinx emblem, and in one corner the Royal Tiger again bares his teeth, balanced in the other by the Arms and Motto of Gibraltar.

All the Battle Honours granted to the old regiments over the years, from Namur 1695 to Korea 1951–52,* have descended to add lustre to the distinctions of the Royal Anglian. To the soldier of today many of these names of long-forgotten battles across five continents may mean nothing; but each represents the soldierly virtues of courage, obedience, loyalty and self-sacrifice displayed by those who have gone before. And so the most treasured are still honoured today by those who now serve: Almanza for the 9th; Blenheim 1704 for the 16th; Minden 1759, when the 12th Foot advanced, bedecked with roses, against the French cavalry, symbolized by the wearing of red and yellow roses on Minden Day by the 1st Battalion; Talavera 1809, when, in the words of the Duke of Wellington, 'the battle was certainly saved by the advance, position and steady conduct

* Those emblazoned on the Colours are listed in Appendix 1.

of the 48th Regiment'; Salamanca, perhaps Wellington's greatest victory, for the 44th; and Sobraon 1846, when the 10th Foot marched like a wall, with bayonets fixed, against the Sikh guns. Each year, on days set aside for the purpose,* the bravery of the officers and men of the old regiments are commemorated by their descendants of the present, so different in many ways, yet in others much the same.

Finally the past is remembered in music. As the Regiment marches past in quick time, it does so to the old march of the Royal Norfolk, 'Rule, Britannia!', added to the more homely country air of the Suffolk's 'Speed the Plough', surely an appropriate tune for a regiment recruited from a still largely rural part of England. In slow time, the music changes to the stately strains of the slow march of the Northamptonshire Regiment.

Having thus established the lineage of the Royal Anglians, it is now time to turn to their accomplishments.

* Almanza, 25 April, 1st Bn; Minden, 1 August, 1st Bn; Sobraon, 10 February, 2nd Bn; Talavera, 27 July, 2nd Bn; Salamanca, 23 July, 3rd Bn; Blenheim, 13 August, 3rd Bn; Royal Tigers Day, 25 June, 4th Bn.

CHAPTER 3

THE GARRISONS

'He carried the sword and the buckler,
He mounted his guard on the Wall'
Rudyard Kipling, *Rimini*.

FORMATION DAY FOUND the four Battalions of the Regiment spread far afield: the 1st Battalion in the Arabian Peninsula, the 2nd at Felixstowe in Suffolk preparing to go to Cyprus, the 3rd in Berlin and the 4th at Watchet in Somerset. As each battalion was to have some share in the last years of British involvement in Aden, this can best be dealt with in a separate chapter, and the narrative will therefore part company with the 1st Battalion for a time, while the activities of the others in their various garrisons are pursued.

3rd Battalion, Berlin, 1964–66

The 3rd Battalion, under the command of Lieutenant-Colonel Norbury, had joined the Berlin Brigade on 14 July, 1964, in its former identity as 3rd East Anglian. The Battalion's operational rôle in the defence of the western sector of Berlin had to be rapidly learned and rehearsed, and the cry of 'Rocking Horse' in the early hours, to turn out the Battalion for a practice manning exercise, soon became familiar to all ranks.

The Reconnaissance Platoon, relieved from time to time by sections from the Rifle companies, carried out regular patrols along the Sector boundary both by day and night, a duty which called for vigilance and fortitude, particularly during the snow and sub-zero temperatures of the winter of 1964. Men of the Battalion who were engaged in these patrols gained a very real understanding of the reason for their presence in Berlin. As the Regimental Journal, *Castle*, put it, they were given:

> 'an insight into the meaning of Communism as they watch the wire and the cleared area beyond it, where East German Communist border guards patrol with their dogs or peer at freedom through binoculars from their watch towers placed at intervals along the wire. There is an eerie feeling as one watches and waits for the next attempt of those seeking freedom; whether they will make it or be shot down in the attempt.'

This longing of many of those in the East to escape was brought home with great immediacy to the wives of the Battalion who, by a coincidence, were able to see the exit of an escape tunnel in a bakery in the Benares Strasse on the very day that the escape took place.

Other duties peculiar to Berlin involved the provision of guards on the Russian War Memorial in the Tiergarten and on Spandau Prison, in which were incarcerated, at that time, the three remaining chief Nazi war criminals—Hess, Speer and von Schirach. The latter task was shared, not only with the other battalions of the Berlin Brigade, but also with the Americans, French and Russians. While this division of responsibility for Spandau ensured that its frequency, for the 3rd Battalion, did not occur too often, the days when the guard was handed over to the other Powers involved considerable protocol and ceremonial, not to mention the interminable prestige lunches that formed an inevitable part of the ceremony. The unique character of Berlin, with its four occupying Powers, each with their own diplomatic and military hierarchy, and the city's own authorities, demanded a continuous round of social activity, which required great stamina for such events from the Commanding Officer and officers of the Battalion and their wives.

Ceremonial, too, played a prominent part in the life of a battalion in Berlin. The 3rd Battalion was heavily committed in the Berlin Tattoo of 1964, an event which saw the greatest concentration of British military bands and drums yet achieved in Berlin. In May, 1965, the Battalion took part in the Allied Forces Day Parade, providing detachments for the Tripartite Colour and Honour Guards, plus the Reconnaissance Platoon mounted in their Ferret scout cars and anti-tank and vehicle mounted troop detachments; the Band and Drums, with the Band of the 1st Battalion Royal Irish Fusiliers, played at the saluting base throughout the parade. Shortly afterwards, on 27 May, the Battalion participated in a Review of the Berlin Brigade, held on the Maifeld to mark the visit to Berlin of HM The Queen and HRH The Duke of Edinburgh. This was a magnificent and memorable occasion, not only for the Battalion and the other British troops, but also for over 100,000 West Berliners, who watched the spectacle and showed their appreciation with cheers. The Allied Forces Day Parade was an annual undertaking, as was the Queen's Birthday Parade and, at the latter event in 1966, the Queen's Colour of the Battalion was trooped before HM Ambassador, Sir Frank Roberts, and a huge crowd, which greeted every drill movement with thunderous applause.

Ceremonial had an importance in Berlin not found elsewhere, as it demonstrated support for the inhabitants of what is, in effect, a beleaguered city in a spectacular, easily-observed yet unprovocative way. An example of this occurred in April, 1965, when the Bundestag of West Germany held a session in West Berlin. On the same day the Battalion was undergoing its Annual Administrative Inspection. As the parade for the Inspection ended, a series of sonic booms were produced by Soviet aircraft, doubtless in disapproval of the Bundestag's meeting. To 'show the flag' and demonstrate support for the rights of the Bundestag, the Battalion, still in the No. 2 Dress worn for the parade, mounted its vehicles and toured the British sector in several columns, a performance that was repeated during the afternoon with the Battalion now dressed for war.

Training areas in Berlin were necessarily restricted and time for it limited, but during its tour there, the Battalion became thoroughly familiar with the Grunewald, a wooded area with a lake within the city limits. Joint exercises with the other Allied troops were a feature of Berlin training and close affiliations were established with the 3rd Battalion 6th United States Infantry and the 3rd Battalion 46th Regiment of the Line (La Tour d'Auvergne) of the French Army. To enable the Battalion to spread its wings more effectively than was possible in the limited area of the Grunewald, annual three-week periods were spent on the Soltau-Luneburg training areas of West Germany, where company and battalion exercises were carried out, culminating in a test exercise set by HQ Berlin Brigade. The first of these, in September, 1964, elicited very favourable comments from the Brigade Commander, and proved the success of the Battalion's training activities and the conversion from a National Service unit to an all-Regular basis that had taken place before arriving in Berlin.

Lieutenant-Colonel Norbury, who had presided over this conversion and the change-over to the large regiment, handed over command on 1 December, 1964, to Lieutenant-Colonel Leng. Among his other achievements, Colonel Norbury had the distinction of having his son, Second-Lieutenant D. M. Norbury, serving in the Battalion with him, a common enough situation in the eighteenth century, but unusual for the 1960s.

Despite its many military commitments, the Battalion found time to participate in all the sporting activities of the garrison, and did what it could to establish good relations with the West Berliners. In sport, the most outstanding successes were in athletics, particularly in the performances of Lance-Corporal Gooden, who became the Inter-Services 100 and 220 yards champion, and who, in the BAOR Individual Championships in 1965, won the 100 and 200 metres with times of 10·7 and 22·1 seconds respectively. On the 'hearts and minds' front, the Band provided countless entertainments at civil gatherings, and the Battalion as a whole adopted the Bezirk* of Spandau, giving parties and other amusements, especially for the elderly and the orphans.

So the Berlin routine of guards and patrols, training and ceremonial went on within this outpost of the West; separated from its other half by the Wall and surrounded on all sides by the dour, suspicious, dedicated face of an alien ideology; inside it, the lights, gaiety, shops, rushing traffic, the tall new buildings of the consumer society with, moving amongst it all, the uniforms of Britain, France and America, the symbols of the West's determination to guarantee the integrity of West Berlin. A remarkable and in many ways rewarding station for a British soldier but of necessity confining, and so, after two years, the 3rd Battalion looked forward to its move in August, 1966, to Tidworth and, as they thought, the different life of the Strategic Reserve with keen anticipation. Their future rôle, however, was to be very different from what they expected.

2nd Battalion, Cyprus, 1964–67

The 2nd Battalion's stay at Felixstowe in the summer of 1964 was short, since it was only staging there between its return from BAOR as 2nd East Anglian, and its departure, in its new identity, to Cyprus. Nevertheless the Battalion managed to fit in a tour of Lincolnshire and Northamptonshire, a visit by HRH The Duchess of Gloucester, and a ceremonial parade taken by Brigadier Oulton to mark the formation of the new regiment.

Cyprus in 1964 was beginning to quieten down again after the outbreak of communal strife in December, 1963. The Battalion's predecessors, the 3rd Green Jackets (The Rifle Brigade), had become heavily involved in these disturbances, following the request by President Makarios for British aid, but, by the end of March, 1964, the task of keeping Greeks

*** District.**

and Turks apart had been taken over by a United Nations Force. By the time the 2nd Battalion arrived in Cyprus in October, it was able to concentrate on its prime task of securing the Eastern Sovereign Base Area of Dhekelia.

Following the grant of independence to Cyprus in 1960, two enclaves on the south coast remained under British sovereignty, each with a military garrison of one battalion. The 2nd Battalion's area contained a number of important installations, while that in the west, at Episkopi, embraced the large RAF base of Akrotiri.

On arrival, the Battalion found it had to take over a considerable number of operational guards on installations around Dhekelia; most were corporal's guards and secured the production of water and electricity on which the base depended, but one, on the Ammunition Sub-Depot, required a complete platoon commanded by an officer, which remained *in situ* for a week at a time. In addition, one platoon was kept permanently on standby at 30 minutes' notice ready to deal with any eventuality. These enterprises were time-consuming and expensive in manpower, but in view of the unsettled state of the island, the need for them was self-evident to all ranks, and the discomfort and long hours were cheerfully borne. However, as time went on and tension in the island was reduced, boredom began to set in and the numbers of men deployed on these tasks seemed hardly justifiable. By dint of constant nagging at HQ Cyprus District, the Commanding Officer, Lieutenant-Colonel Chambers, managed to get the guards reduced to more realistic proportions, thus sparing more time and men for other matters.

Another contingency task the Battalion faced, but mercifully never had to undertake, was the evacuation into the Sovereign Base Area of British Service families from Famagusta, on the east coast. In the curious way things are sometimes arranged by the Staff, none of the families in Famagusta were the Battalion's, whose dependents, if not living in Dhekelia, were in Larnaca to the west; the Larnaca families, however, were the responsibility of another unit, whose own families doubtless lived in Famagusta! The enormously complex problem posed by this masterpiece of planning nevertheless had to be studied and plans worked out and rehearsed.

In addition to its duties in the Sovereign Base Area, the Battalion had another rôle, that of an airportable reserve for rapid deployment outside the island. The problems of air-loading had to be resolved and all training in Cyprus had a counter-insurgency setting to fit the Battalion for such a rôle. Training grounds within the Base Area were limited and did not offer much scope, but between Dhekelia and Larnaca was a strange, barren, almost lunar landscape known by the whimsical name of Goshi-Troulli, which afforded sufficient space to exercise the whole battalion. At the other end of the island was an ideal area for counter-insurgency training, known as Evdhimou, which was covered in trees and undergrowth and seamed with gullies. These two locations, together with some smaller areas, notably the idyllic Akamas peninsula in the north-west, offered a good variety of terrain, plus facilities for field firing, and many interesting and realistic exercises were enjoyed therein.

By mid-1965 the situation in the island had eased to the extent whereby it became possible to start training further afield, and in September, 'C' Company acquired their first experience of desert conditions with a fortnight in Libya. They were followed by 'A' Company, and in March, 1966, the whole Battalion spent three weeks there, repeating the operation the following year. The first sight of the maps, issued for use in Libya, proved a little daunting since they showed the grid squares and very little else. However, the problems of desert navigation were studied apace and once officers and NCOs had gained confidence in the use of the sun compass, success in crossing acres of featureless desert, either by day or night, was soon achieved.

The first trip to Libya gave Lieutenant-Colonel Haycraft his first opportunity of handling the Battalion in the field, away from the varied pressures in Dhekelia. He had assumed command in November, 1965, from Lieutenant-Colonel Chambers, who had ended his tenure with a fine parade, at which the Regimental Colour, presented to 2nd East Anglian in July, 1963, was trooped for the first time. This parade also provided the Battalion with an opportunity to show its appreciation to Brigadier Oulton, who was about to hand over his Deputy Colonelcy to Brigadier Green. The Battalion had had little time for any ceremonial since arriving in Cyprus, so to raise the drill to the standard required and to choose the company which was to provide the Escort, an inter-company drill competition was held. This was won by a photo-finish by 'A' Company from HQ Company, a result which reflected great credit on the Company Sergeant-Major, WO 2 Bird.

Sporting facilities in Cyprus were plentiful and within easy reach of Alexander Barracks. Two of the most successful teams were the Battalion Basketball team which was consistently undefeated and 'B' Company Football team which, in the 1965–66 Minor Unit Championships, won both the league and the knock-out competitions. Having a barracks close to the sea ensured a large number of competent swimmers and in 1965 the Battalion won the Cyprus and World-Wide Life Saving competitions. The proximity of the Troodos mountains offered skiing, at which the Battalion teams were so successful that, in the Inter-Services Competition of 1967, the Army team consisted entirely of 2nd Battalion skiers, a feature of their achievement being the outstanding performance of Lance-Corporal Hardy.

Since the war equestrian activities have been largely the preserve of the Cavalry, but in Cyprus, under the aegis of Captains Taunton and White, the Battalion ran the Garrison Saddle Club and the Dhekelia Drag Hunt. It also formed a polo team, two members of which, Privates Crowston and Tompkins, had had little or no acquaintance with the horse before going to Cyprus.

At shooting the Battalion scored an astonishing, almost embarrassing win over the 1st Battalion The York and Lancaster Regiment in an Inter-Battalion Match for platoon teams in November, 1965. Out of 24 teams taking part, 2nd Battalion teams took the first nine places, 12th, 14th and 16th! This victory represented the fruits of much hard work by the Weapon Training Warrant Officer, WO 2 Dixon. The following year the Battalion did it again, winning all matches and gaining several individual prizes.

A distinction the Battalion enjoyed at this time was its Air Platoon. In 1965 a number of regiments and battalions had their own air platoons, but by September, the 2nd Battalion's was the first in the Army to have its full establishment of Regimental pilots: Major Smith and Captains Tyzack and Everitt. As someone wrote in *Castle*: 'A shiny all Royal Anglian chopper squatting on the square with its all Royal Anglian crew lounging nonchalantly beside it makes us all feel very up to date'. Initially equipped with Austers, the Platoon got Sioux helicopters in 1966. Unfortunately the novelty of these machines made them so much in demand outside the Battalion that the Battalion's connection with its own Air Platoon seemed to be confined to the administration of its members.

Not long after the Battalion's arrival in Cyprus prompt and courageous action in extinguishing a dangerous fire just above the underground petrol tanks by three members of the MT Section was recognized by the award of the Queen's Commendation for Brave Conduct to Corporal Stokes REME and Privates West and Stamps. In December, 1965, Corporals Kirman and Stratford and Private Bannister received a GOC's Commendation for Courageous Conduct for their action when a truckload of white phosphorous mortar bombs overturned. Another fire, in the guardroom of one of the many external guards undertaken by the Battalion (in an empty camp), cost the lives of Corporal Jackson and Lance-Corporal Boothright of 'A' Company, together with those of four Cypriot policemen.

The Battalion had been due to return home in June, 1967, but the move was delayed by the outbreak of the Six-Day War. However, by September the Battalion was back at

1. An inheritance from the 9th Foot; this magnificent silver centre-piece is now in the Officers' Mess of the 1st Battalion.

The Queen's and Regimental Colours

REGIMENTAL COLOUR.

of the 1st Battalion, The Royal Anglian Regiment.

4. A patrol from the 3rd Battalion in Sheikh Othman, 1967.

5. A mechanized patrol of the 1st Battalion in the rugged Radfan country in 1964.

Felixstowe as part of the Strategic Reserve. Nearly three years had been spent in Cyprus and most of the single men were probably glad to leave. For the married men and their families, the island had much to offer: the sunshine, the long, almost empty, sandy beaches, the quintessential Mediterranean landscape of olive trees and orange groves, the pine-scented mountains affording welcome cool from the sun-baked plain in summer, ancient remains, and cheap and plentiful wine. All these and other delights, on which tourists, then just beginning to appreciate the possibilities of Cyprus, were prepared to expend considerable sums of money, all this was freely available to a garrison battalion. Guards and duties proved onerous and sometimes unnecessary, particularly early on, but officers and men always had a fine barracks by the shore to return to, and plentiful opportunities for sport and training provided respite from the long hours spent on sentry in tall watchtowers or sand-bagged bunkers. In short, a not unattractive sojourn by the sea for the 2nd Battalion.

4th Battalion, Watchet, 1964; Malta, 1965–68

By 1964 Doniford Camp, Watchet was on the way to becoming an Anglian preserve. The 4th Battalion had arrived back there in March, as 1st Royal Leicesters, from a successful operational tour in Borneo where they had previously served in 1962–63. It was also familiar to many of the 2nd Battalion, who had been there in the late 1950s as Northamptons and again in the early 'sixties as 2nd East Anglian.

The 4th Battalion spent most of the year training for their new rôle in the Strategic Reserve, and broadening the military minds of many of the young officers, NCOs and men, who, though seasoned jungle warriors, had no experience of other military skills. Shortly after Formation Day a party spent three weeks touring Leicestershire and Rutland, accustoming the inhabitants to the Battalion's new identity. Then, early in 1965, the Battalion was suddenly warned for a six-month tour in Aden, where the 1st Battalion was still serving. A rear party and the families were to remain in Watchet, but all were told that they would have an accompanied tour in Malta to look forward to, starting in November.

The Aden experience, like those of the other battalions, will be covered in the next chapter, so at this point the narrative will move ahead to the end of 1965 and the Battalion's arrival in Malta.

As the Battalion's correspondent to *Castle* put it: 'We expected to find a quiet military backwater where nothing much happened, and where we could take it easy for a bit after the rigours of Aden. Alas, we were quickly disillusioned, soldiering here seems as hectic as anywhere else.' The Battalion's rôle in Malta was to assist in the protection of British interests on the island – such as guarding Vulnerable Points – should the internal situation require it, and to carry out similar duties in Libya where, at that time, there were still British troops in Benghazi and Tobruk. The Battalion's responsibility lay in Benghazi, where a detachment, which varied from a platoon to a company, was provided on three-monthly rotation from Malta. In addition, the whole Battalion was to be prepared to move to Libya for operations in the event of an external threat to that country, at that time still ruled by a monarch friendly to Britain.

The Battalion was quartered in St Patrick's Barracks by the sea on the north-east coast of Malta, with one company detached at a camp at Ghain Tuffieha, some 11 miles away. Life in Malta resembled, in the words of an officer who served there, 'Garrison duties in the old colonial style; fun and relaxing at first but with an increasing need, as time passed, to guard against enervation and boredom, particularly as training areas were so limited, and the inevitable garrison chores, fatigues and guards reduced the numbers of men available.'

The proximity of Malta to North Africa afforded the Battalion opportunities for training in the Libyan desert. The aim was to get each company away to Cyrenaica for a three-week period whenever this could be arranged, and two battalion exercises were held there annually. Like the 2nd Battalion, the 4th soon became adept at living and moving in the desert and these ventures proved a welcome break from the routine of life in Malta. One such exercise, in September, 1966, gave the Battalion an insight into the workings of the United States Marine Corps. The exercise involved an opposed landing by the 6th Battalion, 2nd Regiment USMC, in which 4th Royal Anglian would defend the beachhead. To add interest to the exercise, it was agreed that each battalion would exchange a company, and it fell to 'C' Company to learn the highly complicated disembarkation drills of the Americans. The whole process was so beset with the jargon beloved by the United States military that at the first practice not one man of 'C' Company arrived at his correct boat station! Nevertheless, for the landing itself the Company managed to deploy itself correctly, despite the operation being carried out in pitch darkness with the ship totally blacked out. The exercise proved instructive and enjoyable, with the Battalion's self-esteem greatly boosted by the Marine Corps' description of them as a unit of 'seasoned veterans'!

The Arab-Israeli War of 1967 severely curtailed the Battalion's ability to train in Libya, though later on exercises up to company level were again permitted, and adventure training expeditions, at one time forbidden entirely, also recommenced.

In March, 1968, Captain Amos led a party of twenty, with seven vehicles, to the Tibesti mountains by way of the Kufra oasis, a well-known base for the Long Range Desert Group during the Second World War. The expedition took three weeks, covered 3,000 miles and experienced every kind of going to be encountered in the desert – gravel plain, huge sand dunes, an area of bright orange sand where the petrol in the 3-tonners vaporized in the fuel lines, and a plateau, described in *Castle* as 'a cross between a gigantic slag heap and a coke tip'. Beyond the latter the party found its way barred by a sheer 300-foot drop on to a sandsea beyond, and though they attempted to find a way round, their fuel situation forced them to turn back without reaching the Tibesti. Despite this, the experience gained in learning how to live and move in the desert was immense and the party had been able to see what an officer described as 'miles and miles of nothing, yet the most stark, awesome and strangely beautiful nothing one can imagine'.

The following month a 'C' Company expedition spent thirty-four days in the desert, driving again to Kufra by way of Gialo, and returning by such landmarks as Lady Be Good, a crashed Liberator bomber from the Second World War. While the arts of existing in the desert in temperatures between 110 and 120 degrees Fahrenheit were being discovered by these expeditions, a party of nine soldiers under Lieutenant Howe were mountaineering in Turkey in temperatures at the other end of the scale. Their aim was to climb Mount Erciyas (3,916 metres). They reached a point 800 feet from the summit, but bad weather, a lack of proper equipment to cope with ice, and the inexperience of the climbers prevented them from making the final bid for the top.

The numbers of men from the whole Battalion who could partake in such expeditions was, of course, limited, but others made a considerable mark in the sporting world. In 1965–66 the Battalion won all the Malta Football, Hockey, Cross-Country, Athletics and Rowing Championships, and scored a notable victory by being first home in the Malta Boat Race, the first Infantry battalion to do so since 1924. This achievement was repeated in 1967, the local papers reporting that 'the Anglians were in a class of their own'. The biggest triumph in 1966 was the Battalion footballers' achievement in reaching the final of the Army Cup. Under the management of Regimental Sergeant-Major Jenks and trained by Sergeant Stoves, APTC, the team won through five matches with convincing scores to meet 24 Signal Regiment in the final at Aldershot. But here, to the bitter disappointment of everyone, the team failed to reach its earlier form and lost 3–0.

Despite this set-back, the Battalion sportsmen continued to do well in Malta, especially in Athletics. Their prowess can be judged by the fact that, in 1968, the twenty-strong Army team for the Inter-Service Championships (Malta) contained nineteen 'Tigers', who gained eleven firsts, nine seconds and three thirds. Corporal Sarson won the Garrison, Inter-Services and Maltese 100 yards and 120 yards hurdles, as well as being the Garrison and Inter-Services long jump champion; Corporal Dutton was the Malta champion in the shot and discus. Both these NCOs, together with Corporals Crooke (220 yards) and Waqairoba (120 yards hurdles) went on to compete with honour in the Army Individual Championships at Aldershot. The most outstanding performance of all was that of Lance-Corporal Holman. In 1967 he defeated all-comers in the High and Triple Jumps and was first string in the Battalion sprints team. In 1968 he improved on his performance in Malta and finished the season at Aldershot as the Army High Jump Champion with a jump of six feet. The Battalion's many successes at sport earned it the comment on its Annual Inspection report for 1967, 'They must be the fittest Battalion in the British Army'.

Fitness in the Battalion began at the top, for the Commanding Officer himself, Lieutenant-Colonel Carter, was a member of the Battalion hockey team. He had assumed command in April, 1966, from Lieutenant-Colonel Cowan, who had seen the Battalion through its conversion to Royal Anglian and commanded during the six-month tour in Aden.

Four months after Colonel Carter's arrival, the Battalion received the Lord Mayor of Leicester, then Councillor Mrs Monica Trotter, who came out to Malta to confer the Freedom of the City of Leicester and to present new trumpet banners. The Lord Mayor, who was accompanied by Major-General Spurling, the Deputy Colonel, spent five days with the Battalion, which gave her an insight into all aspects of garrison life on the island. During the visit to Malta in November, 1967, of HM The Queen and HRH The Duke of

Edinburgh, the Battalion played a large part in the Searchlight Tattoo, the climax of which was a modern battle scene devised by Major Robertson. Another Royal visit took place in May, 1968, when HRH The Duchess of Gloucester spent two days with the Battalion. A month later the Battalion again entertained the Lord Mayor of Leicester, now Alderman Sir Mark Henig, and his Lady Mayoress, who spent five days seeing as much of the soldiers' life as possible.

The Benghazi commitment finally came to an end in February, 1968. Lieutenant Pepper's platoon was the last to leave and their return to Malta was greeted with some ceremony, including a fly-past of Canberras! Apart from a two-month tour in Aden by 'C' Company, which will be referred to in the next chapter, Benghazi had provided the only chance of 'real soldiering' for men of the 4th Battalion during their stay in Malta. On Monday, 5 June, 1967, the Arab-Israeli War erupted. By 10.00 hours, 4 Platoon 'A' Company had been alerted to protect the British Consulate-General and the American Consulate. Several thousand rioters had already paid their call on the Consulate, burning cars and breaking into the ground floor, moving on to the Consulate-General to smash windows and set fire to the Residency flats, before going back to threaten the Americans again. By the time the rioters returned to the British offices for the second time, 4 Platoon were in position, and the crowds contented themselves by venting their frustrations on cars parked in the vicinity. The Platoon then had the task, with a troop of the 5th Royal Inniskilling Dragoon Guards, of escorting the staffs of the Consulates to the sanctuary of D'Aosta Barracks.

During the night of 5–6 June 4 Platoon were out again, escorting families to Wavell and D'Aosta Barracks through streets where fires and rubble marked the sites of houses or shops owned by Jews or Americans. By 06.00 hours on 6 June 2,000 people had been concentrated in the safety of the barracks.

That afternoon the remainder of 'A' Company, who had

been firing their range classification in Malta, flew into Benina airport, to face a somewhat delicate confrontation with some Algerian paratroops on their way to the East to demonstrate Arab solidarity. Anglians and Algerians gazed at each other with some curiosity for some time, but eventually the latter flew off and 'A' Company were smuggled away to Wavell Barracks. Here they were joined by 4 Platoon, and the Company settled down to four weeks in cramped quarters, helping the Inniskillings with the evacuated families and on the alert for any further eventuality until the tension diminished.

The situation remained uncertain until November and, with large numbers of British civilians in the city to protect, a company was kept permanently at Wavell Barracks, changing over every six weeks. In December the closing-down process began. Wavell Barracks was taken over by the Libyans on 19 January, 1968, and on the 31st a guard, found by 5 Platoon, witnessed the final hand-over of D'Aosta Barracks and the end of the British garrison in Benghazi.

Five months later, in early July, the 4th Battalion left Malta for its next station, at Gillingham, and its new rôle in the Strategic Reserve.

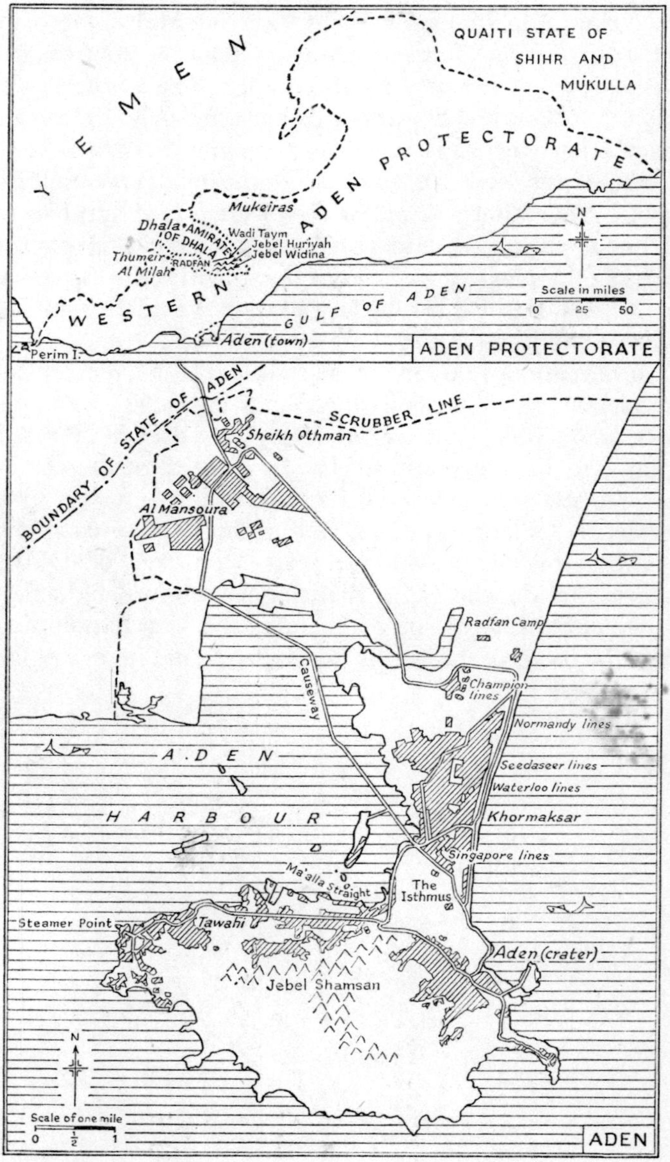

YEMEN
QUAITI STATE OF
SHIHR AND
MUKULLA
WESTERN ADEN PROTECTORATE
Mukeiras
Dhala
AMIRATE OF DHALA
Wadi Taym
Jebel Huriyah
Jebel Widina
Thumeir
Al Milah
RADFAN
GULF OF ADEN
Scale in miles
0 25 50
N
Aden (town)
Perim I.
ADEN PROTECTORATE
BOUNDARY OF STATE OF ADEN
SCRUBBER LINE
Sheikh Othman
Al Mansoura
Radfan Camp
Causeway
Champion lines
Normandy lines
Seedaseer lines
Waterloo lines
Khormaksar
Singapore lines
ADEN
HARBOUR
Ma'alla Straight
The Isthmus
Steamer Point
Tawahi
Aden (crater)
Jebel Shamsan
N
Scale of one mile
0 ½ 1
ADEN

CHAPTER 4

THE BARREN ROCKS OF ADEN

'When first under fire an' you're wishful to duck
Don't look nor take 'eed at the man that is struck.
Be thankful you're livin', and trust to your luck
And march to your front like a soldier.'
Rudyard Kipling, *The Young British Soldier*.

IN THE LATE 1950s that part of the Army's world known generically as 'Aden' consisted of a Crown Colony, which included the port and strategic staging post of Aden itself, the BP oil refinery five miles to the west, and, to the north and east, a number of tribal territories divided into the Western and Eastern Aden Protectorates. The latter will not concern us here but the Western Protectorate, made up of eighteen amirates and sheikhdoms, formed a buffer between the Colony and the aggressive state of the Yemen, which seized every opportunity to fester discontent and rebellion among the unruly tribes of the Protectorate.

In 1959–60 the British Government formed the Protectorate states into the Federation of South Arabia, to which, in January, 1963, was joined Aden Colony itself, much against the will of its inhabitants, whose nationalism, fired by ceaseless propaganda from Egypt, was fiercely opposed to the feudal and poverty-striken states of the Protectorate. The passions besetting this union had been further inflamed by the establishment in September, 1962, of a Republican régime in the Yemen, supported by Egyptian troops. While these political events were taking place, Britain's military requirements in the Middle East were transforming Aden from a staging post into a major base, with the setting-up of Middle East Command in 1961 and the transfer, from Kenya, of the brigade forming the strategic reserve for the Middle East; all of which added fuel to the fires of Arab nationalism in the Arabian Peninsula.

The first manifestation of the serious trouble that was to ensue occurred, not in Aden itself, but in the hinterland, in the mountainous area known as the Radfan, through which the road from Aden passed to the Amirate of Dhala, 80 miles to the north, where there was a British garrison close to the Yemen frontier. The Dhala area had always been turbulent and movement of convoys to it required a two-day operation with armoured car and infantry escorts and the hills along the route secured by picquets. For those of a romantic turn of mind, service in the Aden hinterland held echoes of the great days of Empire, the years of skirmish and ambush on the North-West Frontier of India; the burning sun, the craggy, scrub-covered hills, the wild tribesmen, each with his rifle, the wheeling hawks, the stone villages, with every house built like a miniature fortress—all this provided a back-cloth of a type familiar to generations of British soldiers. In 1964, amid the towering peaks and fertile wadis of the Radfan, British troops once more embarked on what was to be the last of this type of 'small war'.

The 1st East Anglian Regiment had arrived in Aden as a garrison battalion in early 1964 and, after providing a company for the protection of the Radfan force's base at Thumeir in April, the whole Battalion, under the command of Lieutenant-Colonel Dye, became involved in the operations in early May. Throughout the month and into June the force steadily drove the tribesmen from the area, until only the final objective of the Force Commander, the Jebel Huriyah, at 5,500 feet the highest peak in the Radfan, remained to be captured. On the night of 10–11 June, after a preliminary attack by the 2nd Battalion Federal Regular Army* on an intermediate position, 1st East Anglian scaled the heights of the great Jebel, their route up the jagged slopes lit by flares dropped by Shackleton bombers, and by 04.50 hours were on the summit, looking down at the lights of Aden town 40 miles to the south. For a week the Battalion held the mountain, until 18 June when it returned to Aden.†

For gallant services during these operations, the BEM was awarded to Sergeant Jones, who took command of his platoon when the platoon commander was wounded, and by his leadership and example, contributed greatly to the platoon's success in its first engagement.

After the capture of Jebel Huriyah the Radfan had to be policed, and in August 'A' Company, under Captain Abbott, carried out a helicopter-borne night assault to seize and clear the Jebel Widina, an operation thought to be the first of its kind in the Army.

Formation Day, on 1 September, 1964, and the months that followed found the 1st Battalion, as it had become, carrying out internal security tasks in Aden, with companies rotating on detachment at Dhala and at another frontier outpost to the east, Mukeiras. In addition the Battalion spent two more periods in the Radfan before the end of the year, and suffered three casualties from a land mine on the last day of 1964: Private Frazer being killed and Corporal Andrews and Private Barrell wounded. On 17 February Lieutenant-Colonel Dye, who had been awarded the OBE for the Battalion's fine work in the Radfan, handed over to Lieutenant-Colonel Creasey. Lieutenant-Colonel Dye returned in December, 1966 as a Brigadier in command of the South Arabian Army, a most demanding appointment which he successfully held until handing over to Aqueed Mohammed Ahmed at the time of the final withdrawal of the British in December.

* Formerly the Aden Protectorate Levies.

† A full account of 1st East Anglian in the Radfan War can be found in *Last Post: Aden 1964–67* by Julian Paget (1969), pp. 51–110.

Though the Battalion continued to find detachments up-country, attention now became focused on Aden itself. The Conservative Government's Defence White Paper of July, 1964, had stated that the military base in Aden would be maintained but that South Arabia would be granted independence not later than 1968. This prompted the Aden nationalists, backed by Egypt, into action to ensure that power would fall into their hands and that the military base was rendered untenable; their methods were to be intimidation and terror. A Labour Government took over in Britain, and a visit by the new Colonial Secretary in November, 1964, was deemed by the chief subversive organization, the National Liberation Front (NLF), to be a suitable opportunity for opening its campaign. By the end of the year, NLF terrorism had caused 36 casualties to the security forces.

The two main groups of the terrorist organization with which the security forces were concerned were the NLF and the Front for the Liberation of Occupied South Yemen, a grandiose title usually abbreviated to a cognomen with old-maidish connotations—FLOSY. The NLF were totally committed to violent methods from the start; FLOSY initially believed in political action but by 1967 had also adopted terror. In 1966 the two joined forces but by the end of the year the NLF had broken away and the following year both groups were not only fighting the security forces but also each other. Though both enjoyed Egyptian aid and backing, FLOSY was committed to Egyptian control after independence, whereas the NLF was not. The terrorists were equipped with grenades, mines, bazookas, small arms and mortars, and made use of all methods of insurrection: subversion and propaganda, strikes and riots, sabotage and assassination. They succeeded, quite early on, in intimidating the population and, by murder and threats, went a long way to neutralizing the Government's sources of intelligence, the Police and the Special Branch.

The arena in which this scene of the grand finale of

Empire was to be played out can only be described as seedy. Though Aden contained a number of modern buildings, they were a façade, a veneer of civilization, behind which festered the narrow, dirty alleys, the squalid, reeking huts and the crumbling, fly-blown houses of the bazaar, teeming with its villainous and unlovely inhabitants. The town of Aden had grown up within and around the crater of an extinct volcano, which was attached to the mainland by a narrow strip called the Isthmus. The predominantly Arab part of the town, known as Crater, was surrounded on the north, west and south by the jagged heights of the volcano's rim, and on the east by the sea. Towering over all was the highest point of the volcano, the Jebel Shamsan. To the west lay the area of Steamer Point, embracing the seat of the Administration, Government House, and the cantonments of HQ Middle East Command. From here the road ran east through Tawahi, which contained the main shopping centre, hotels, the port and a large Arab quarter; then along the mile-long Ma'alla Straight, which had a large number of flats housing Service families, with a maze of Arab dwellings behind them. Beyond Ma'alla the road forked: to the right it went through a pass in the volcano's rim into Crater; to the left and north, it led up across the Isthmus. Here, in the Khormaksar area, lay the various camps and barracks: from south to north, Singapore, Waterloo, Seedaseer and Normandy Lines. Beyond the latter was the RAF and civil airfield, and further north still, at the top end of the Isthmus, Champion Lines and Radfan Camp.

The road then led off across the desert towards Dhala and the Yemen, but as it left the northern boundary of Aden State it passed through the purely Arab town of Sheikh Othman, with, to the west, the quarter of Al Mansoura, which contained the Detention Centre for suspected terrorists. Across the desert to the west lay Little Aden with the Oil Refinery and the cantonments of 24 Infantry Brigade.

Responsibility for internal security rested with the Aden Brigade, of which the 1st Battalion, stationed in Waterloo

Lines, formed part, together with, in early 1965, 1st Battalion The Royal Scots, who were relieved in Radfan Camp in February by the 4th Battalion, as mentioned in the last chapter. In April, the Brigade was joined by 1st Battalion The Royal Sussex Regiment.

As the terrorist campaign began to build up in 1965, the 1st Battalion had responsibilities all over Aden for the first three months—manning joint military/police headquarters, providing foot patrols, four pairs of mobile patrols, a platoon at immediate notice, and cordons and searches of highly insalubrious areas in Crater and Sheikh Othman. During February several Battalion patrols came under grenade attacks; as a result of his action in one of these Corporal Day was awarded the Queen's Commendation for Gallantry. On 12 March a Blindicide rocket was fired at Sheikh Othman police station, exploding in a room occupied by 7 Platoon headquarters of 'B' Company, seriously wounding the Platoon Sergeant, Sergeant Smith, and destroying the wireless and telephone. Private Elba-Porter, the platoon signaller, was wounded in the eyes and shoulder but, though temporarily blinded in one eye, he fetched another wireless and continued to pass orders so that effective countermeasures could be taken. Private Kent saved the life of Sergeant Smith by rendering immediate first aid, keeping him alive until help came. Both men were awarded the BEM for Gallantry and the Platoon Commander, Second-Lieutenant Copping, the Queen's Commendation for Brave Conduct.

On 2 April the Battalion's area of responsibility was defined as Crater, Khormaksar and Sheikh Othman. As the terror built up, so the duties increased. One night in bed in three became the norm and that night often meant going off-duty at 23.00 hours, in bed at midnight, and on duty again at 07.00 the next morning. To keep the seething alleys of Crater quiet, a company was deployed there each night with assault pioneers, mobile searchlights and Land-Rover patrols. This force had to find two road blocks on the two entrances into Crater, observation posts on roof tops, foot patrols in the streets, and a reserve to deal with incidents as they occurred. By day, road blocks, vehicle checks and mobile patrols were found by this company. Another company was usually committed to static guards, and the third, if available, formed the Battalion reserve. Between February and September, in Crater, 35,000 locals were searched and 8,000 vehicles for a yield of twelve grenades, one mine, six pistols and a few oddments; the deterrent value of such searches cannot, however, be measured. In addition to the more conventional operations, the Battalion formed a Special Branch squad of seven under Captain Horrex which was constantly engaged on clandestine tasks, for which all its members received a Commander-in-Chief's Commendation.

As a result of the terrorists' success in intimidating the population and subverting the Police, the acquiring of information became increasingly difficult. On 19 June, 1965, men of the Battalion achieved a rare break-through. At about 7.30 pm two terrorists threw a grenade into the Seamen's Mission at Steamer Point. A few seconds later, Second-Lieutenant Hawkins, who was returning in a Land-Rover from visiting the Government House Guard, saw three men leap into a car and drive off. He told his driver, Private Richardson, to give chase. Richardson drove with great skill to keep up with the Arab car through the narrow, twisting back streets of Tawahi, until the fugitives drove up a cul-de-sac. Cornered, the three men jumped out. Hawkins managed to grab one of them and brought him down. He then sent Richardson off to fetch help while he sat on his prisoner's chest with revolver drawn, amid a rapidly gathering and increasingly hostile crowd. He could easily have been overpowered but he held his ground until relief, summoned by Richardson, arrived. From the prisoner and the car a great deal of information was obtained, which resulted in many other arrests and the break-up of a complete terrorist cell for six months. Second-Lieutenant Hawkins had not only shown presence of mind but great bravery, particularly since the terrorists were armed and troops were only allowed to use their weapons in self-defence, but his reward could only be what one historian has called 'an incongruous MBE',* since troops in Aden were not officially on active service. Private Richardson received the Queen's Commendation for Brave Conduct.

Shortly after this incident the Battalion had a break from Aden, returning to the Radfan for the last time from 28 July to 22 August. The Battalion positions came under attack twice from machine-gun and rocket fire and, on 15 August, 'C' Company mounted a four-day patrol into the Wadi Taym which resulted in the destruction of an enemy base.

Returning to Aden the Battalion found itself plunged into intensive operations, following the murder in Crater of a British Police Superintendent and the Speaker of the Legislative Council. A dawn to dusk curfew was enforced on 2 September, nineteen days before the Battalion was due to leave Aden. The number of men this required, plus other duties, together with all the arrangements for the move, imposed a great burden on the Battalion; 'C' Company, for instance, spent their last four days in Aden on continuous duty. Finally, on 21 September, the Battalion handed over its duties to 1st Battalion The Prince of Wales's Own Regiment of Yorkshire, the companies moving straight from their operational tasks on to the aircraft.

The 1st Battalion had had an extremely arduous tour in Aden, being almost constantly on operations, which, as the GOC said in his farewell address, had given them a knowledge and experience of internal security duties second to none. From these duties, they went to a very different rôle, that of a mechanized battalion in BAOR, where they will be met again in Chapter 7, but a recognition of their prowess in Aden came the following year with the award of the OBE to Lieutenant-Colonel Creasey in the Birthday Honours.

Meanwhile the 4th Battalion had left Aden before the 1st, having gone out only for a six-month tour from Watchet in

* Gregory Blaxland, *The Regiments Depart* (1971), p. 449.

February, 1965. Whereas the 1st Battalion had been able to rest from their labours in the modern barracks of Waterloo Lines, the 4th, by virtue of their temporary sojourn, had to endure the rigours of Radfan Camp, a euphemism for the sand-blown tents of the transit camp.

On 15 February, three weeks after arriving, the 4th Battalion moved up to the Radfan proper, and for two weeks enjoyed the novelty of picquets and patrols. Then it was back to Aden to take over the area west of the 1st Battalion in Crater. Their duties were similar, companies rotating every three or five days between guards and internal security tasks, with one company detached up-country.

In June 'B' Company came under command of 2nd Battalion Coldstream Guards in the Radfan. During an ambush action, Private Ryder was severely wounded by a grenade, but thanks to the skill of the Company RAMC orderly and helicopter evacuation to Aden, he was on his feet in five days. Shortly after the Company's return to Aden, it carried out a very successful cordon and search which divulged a quantity of ammunition and explosives. The information which led to this operation came from the 1st Battalion, so it was very much a Regimental success.

With two of its battalions in the same brigade, the significance of the 'Large Regiment', then a novel concept, was soon made clear to the rest of the Army in Aden. As a correspondent to *Castle* wrote: 'In conversation we were always referred to simply as "the 1st Battalion" or "the 4th Battalion". There was no need to mention the Regiment.'

In addition to their duties in Aden or the Radfan, platoons and companies also took their turn at finding garrisons for other up-country locations: Al Milah, to protect sappers engaged on road maintenance; Perim Island, 200 miles west of Aden, and Mukeiras, all of which the 1st Battalion had also experienced. The 4th Battalion also managed, despite the troubled times, to maintain its sporting reputation by winning the Middle East Land Forces Soccer Cup.

From the beginning of July through to August, the

companies began to move back to England as the relieving battalion started to arrive. On 16 August 9 Platoon was waiting, in plain clothes, all weapons and equipment packed, ready to fly home next day. At 7.30 pm three incidents occurred in Aden. Within 45 minutes the Platoon was in uniform, armed and fully equipped, sitting in their vehicles ready to move. Such were the hazards of life in Aden in 1965.

The following year was to see a marked increase in pressure on the security forces. The Defence White Paper of February, 1966, not only confirmed independence by 1968, but also announced a complete withdrawal of all forces by that date. Politically, for the Federal rulers, who had always believed in British support for their sovereignty, this was seen as betrayal. Militarily, it meant that the security forces could no longer hope for any local support in their struggle against the terrorists; no Arab was going to put his survival after 1968 at risk by aiding the British in 1966. For the terrorists the news was a godsend and their activities increased, in quality as well as quantity.

The performance of the bomb throwers—known as Cairo Grenadiers—became more effective, and the mortar began to replace the rocket launcher as a favourite terrorist weapon. Sabotage and murder were stepped up. The primary means for curbing terrorism was the search for suspected terrorists and weapons. One measure to restrict and check movement into Aden was the construction of an 11-mile long wire obstacle across the desert from coast to coast, just north of Sheikh Othman, known as 'The Scrubber Line'; all traffic coming down from the Yemen and the Federation now had to enter Aden through one of four check points.

After the departure of the 1st Battalion in September, 1965, there was no Royal Anglian presence in Aden until May, 1966, when 'B' Company of the 2nd Battalion arrived from Cyprus for two months security duties around Khormaksar airfield. Writing in *Castle* of this episode, a member of the Company vividly portrayed a predicament common to the security forces, governed as they were in the face of increasingly ruthless terrorism by the principles of minimum force and the use of firearms only in self-defence:

> 'The back of a 3 tonner. It's dark and we are waiting to go. "It's like this as I see it. This chap comes at you with this bomb, you see. He throws it and then runs. You've got your rifle already cocked, people everywhere . . . the bomb goes off . . . you can see him running through the crowds . . . you can't shoot, can you? . . . I mean, can you? Bloody dangerous this 7.62 stuff . . . goodness knows where it would end up".'

Soon it was the 3rd Battalion's turn. Arriving from Tidworth in October under the command of Lieutenant-Colonel Leng, it moved into Radfan Camp for a nine-month tour and took over responsibility for Sheikh Othman and Al Mansoura. This was probably the most unpleasant and most crucial area in Aden at that time, and the terrorists were keen to control it. The main road to the hinterland, down which came men, arms and ammunition from the Yemen, ran through it. It could not be sealed off, as Crater could be, by blocking the two entrances. Since Crater was regarded as the real capital of Aden by the nationalists, the disturbances there were of a less destructive kind. Sheikh Othman, on the other hand, became increasingly the scene of a trial of strength between the terrorists and the security forces, the former using small arms, mortars, rockets, grenades and mines in their efforts to defeat the troops in open street warfare.

The 3rd Battalion soon began to get a taste of what was in store for it. On 3 November a mine blew up a Land-Rover carrying the Commanding Officer, who was shortly to hand over to Lieutenant-Colonel Dymoke; fortunately there were no casualties. A grenade attack on a patrol on 20 November seriously injured a soldier. Sergeant Allan of the Battalion Medical Team heard the contact report on the radio and, without an escort, drove to the scene of the incident and gave first aid to the wounded soldier. Sergeant Allan later received the BEM for gallantry and the citation recorded that his 'skill and initiative undoubtedly contributed to saving the soldier's life'. On 3 December Captain Cox, RAMC, the Battalion Medical Officer, reached a wounded man within three minutes of a grenade attack and, while giving first aid, another grenade exploded only five yards away. Undeterred by this, Captain Cox continued treating the wounded; he was subsequently awarded the MBE for gallantry. Two days later Radfan Camp came under fire from drainpipe mortars, but of the thirty bombs the terrorists planned to let off, only five exploded inside the camp perimeter.

The Battalion was not slow to score a few successes against the terrorists. Corporal Jarvis captured a suspect digging holes in the road, which led to more arrests and the break-up of a mine-laying gang. On 30 December a quantity of arms and explosives was found in a house in Sheikh Othman. This coup was the work of the Battalion's Special Branch Squad, formed from ten men of the Reconnaissance Platoon under Captain Light. This squad operated with speed and surprise, mostly at night, and though they became prime targets for the terrorists, they were only attacked twice; the first attempt resulted in the death of a Cairo Grenadier, killed by Private Bourke, and on the second Sergeant Hutchinson ambushed and captured three terrorists before they could make their attack. By the end of the Battalion's tour, the Special Branch Squad had captured 105 grenades, five automatic weapons, three pistols, two rocket launchers, a quantity of bombs, ammunition and explosives, and fourteen terrorists, including the second-in-command of FLOSY and the leader of FLOSY in Crater.

As the tempo of terrorism stepped up in 1967, the Civil Police, to whom the Army had hitherto been in support, became increasingly less effective. Early in the year, therefore, the High Commissioner decided that, in future, the Army was to take over primary responsibility for internal security, with the Police in support. This meant that, instead of waiting for a call for assistance to disperse illegal assemblies, the Army was to take the initiative to prevent crowds gathering.

11 February, the anniversary of the formation of the Federation, was designated by the NLF and FLOSY, as 'the Day of the Volcano'. However, the volcano failed to erupt, for, in pursuance of the new policy, the troops deployed the day before and a curfew was imposed.

In April came the fruitless visit of the UN Mission to Aden. The strange behaviour of its three members produced moments of pure farce, though the humour of the situation was probably none too evident to the 3rd Battalion, who bore the brunt of the ensuing terrorist activity. On the day before the Mission arrived—appropriately it was April Fool's Day—Aden endured one of its rare rainstorms, in which six inches of rain fell in seven hours. Streets were flooded, everything came to a halt, and troops spent the day helping to clear up the sodden mess. Small thanks they got, for the next day the NLF called a strike, which brought crowds into the streets, giving cover for the terrorists to throw grenades and snipe at the troops. The next day the tempo increased, particularly in Sheikh Othman. As *Castle* put it: ' "Disperse or We Fire" banners were hastily rolled up and street fighting tactics became the order of the day'. The Police Station was besieged by mobs and had to be relieved by the 3rd Battalion and armoured cars of The Queen's Dragoon Guards. After a quieter day on 4 April, trouble again exploded on the 5th, when the Mission had to be escorted to the Al Mansoura Detention Centre, a visit which provoked pandemonium within the Centre, and a blazing gun battle without. Finally, on 7 April, to everyone's relief, the Mission left Aden, and the terrorists, who had used this visit as an opportunity to try to take over Sheikh Othman, withdrew.

As a result of their work during this turbulent week, Lieutenant Harrington-Spier and Corporal Valentine received the MBE and BEM respectively. The action in which they won their awards, typical of the Aden operations, was graphically described by Corporal Valentine in *Castle*:

'Grenade', the shout that had so many times sent men of the 3rd Battalion diving to the ground was now pounding in my ears. I turned quickly to see Pte. Anderson appear from the smoke of the explosion and the grenadier making his bid for escape towards the Main Mosque. My mind racing, I let off two quick shots and yelled for pursuit. Before the section had gone 50 yards, I was shouting at them to take cover from the heavy automatic fire which was being directed at us from roof tops across the street. The firing stopped, we had silenced the guns, or more likely they had served their purpose. Our wounded grenadier had escaped to the Mosque, which we were forbidden to enter.

The date—6 April, 1967. Aden was in the grip of a General Strike, specially laid on to greet the UN Mission. The time—0900 hrs, and 13 Platoon, 'D' Company, commanded by Lt. Harrington-Spier, was moving into Section 'B' of Sheikh Othman in two patrols. Mission—to detect terrorist activity—mines, mortars, and generally to get a feeling of the strike-bound town. In the preceding two days there had been many shootings; energa and hand grenade attacks, but no riots.

I quickly checked Pte. Anderson and found him to be only slightly injured from grenade splinters. The order 'continue the patrol' followed my sitrep. Once again the section pushed forward, moving steadily and slowly, each man covering the next as we slipped from doorway to doorway right along the front of the Mosque and then across the road into Street 5. Only half of the section were across the road when all hell broke loose from terrorist positions at the end of the street, in the Mosque itself and from tall buildings on either side of us. I heard a heavy exchange of fire from the Mosque behind me and realized that my 2IC, L/Cpl. Jarvis, and Ptes. Donald, Theophile and Butcher had been pinned down. Further back the platoon commander and his patrol had come under fire from the positions we had encountered earlier in the grenade incident. Fire was also coming from the streets to my left as Sgt. Green's patrol tried to work their way round to lend us support. Pte. Lanaghan and I covered each other as we moved in short dashes to the end of the street, searching for a way out. There seemed little hope at present, for on rounding the corner Lanaghan hurriedly dived back, as if blown by a mighty gust of wind!

Returning along the street, I checked on ammunition and casualties ready for the next move. The fire had not lessened. It seemed that the terrorists had plenty of ammunition for their Russian-made automatic rifles.

'Reorg. on Street 10.' The order came from the platoon commander, with the news that a Saracen APC and Saladin armoured car would be covering our move. We regrouped at the end of Street 5. Counting the men, I found two were missing, Theophile and Butcher. Lt. Harrington-Spier, with Lanaghan, McCormick and Anderson moved back down the street to help them. They were still pinned down opposite the Mosque. A grenade came from the top of one of the tall buildings and landed not three feet from McCormick, giving him a chance to exercise his lungs. He carried on shouting 'grenade' until it did what was expected and went off! Luckily—no casualties. Theophile and Butcher were retrieved and we started the move out step by step, from doorway to doorway, corner to corner, down to Street 11 to behind the cattle sheds where Company HQ and 14 Platoon were in position. We dashed the last 200 metres across open ground, and the patrol was out. It was over. We could relax a little.

In addition to the perils and excitements of keeping order in Sheikh Othman, the Battalion had responsibilities for manning check points on the Scrubber Line, a monotonous duty frequently enlivened by terrorist mortar and rocket attacks. On one occasion, Check Point Golf, then manned by the Corps of Drums, came under fire for forty-five minutes. The Drummers held their positions and Sergeant Sutcliffe and Lance-Corporal Ford received Commander-in-Chief's Commendations for bravery.

The Royal Anglian presence in Aden had been increased in early 1967 by the arrival of 'C' Company of the 4th Battalion, who came for a two-month tour from Malta to take over security duties at RAF Khormaksar, the same rôle that the 2nd Battalion company had undertaken the previous year.

On 24 May the Royal Anglian's part in the Aden emergency came to an end, when the 3rd Battalion handed over Sheikh Othman to 1st Battalion The Parachute Regiment. In six months the Battalion had had 459 incidents and 174 casualties, fortunately none of them fatal.

The Aden experience gave the Royal Anglian Regiment its baptism of fire. In so doing, it proved that the formation of the Regiment, considered in some quarters as purely an administrative convenience, had in no way been detrimental to the qualities required by infantrymen on active service. In *The Regiments Depart*, Gregory Blaxland, himself an ex-infantryman, wrote: 'The intimate pride of belonging to a regiment can be a powerful support when under ordeal by fire and steel'; after quoting examples from Korea, Malaya and Cyprus, he went on: 'It was to be seen again in Aden, displayed both by old regiments and those recently born of amalgamation, such as the Royal Anglians.' Perhaps it might be said that, as the holocaust of the Great War finally welded together the old county regiments, formed from the numbered regiments in 1881, so the bullets and bombs of the Radfan, Crater and Sheikh Othman helped to cement the 1964 union of four regiments into a living entity, recognized and acknowledged throughout the Army.

CHAPTER 5

A SIGHT FOR THE QUEEN

'We'll proudly point to every one
Of England's soldiers of the Queen.'
L. Stuart, *Soldiers of the Queen.*

At a Tattoo in Colchester in 1968, as the massed bands and drums of three infantry regiments crashed out the final bars of 'Under the Double Eagle', the commentator invited the audience's applause with the words: 'The Queen's Division, ladies and gentlemen, a sight for the Queen!' In so doing he drew their attention, albeit obliquely, to yet a further reorganization of the Infantry, the genesis of which had occurred the year before.

By the middle of 1967 the battalions of the Royal Anglian Regiment could look back on nearly three years of soldiering under their new designation; fairly uneventful years for the 2nd Battalion, but for the 1st, 3rd and 4th, a period high-lighted by the rigours and dangers of the campaign in the Arabian Peninsula. The mental adjustments necessary to embrace whole-heartedly the conversion to Royal Anglian had obviously been harder for the 4th Battalion than for the other three, whose brief lives under the guise of 1st, 2nd and 3rd East Anglian had never been entirely satisfactory; such nostalgia as existed hankered more for the old county identities than for these sometimes uneasy unions. On the whole though, the battalions looked forward to the future, relatively content with their transformation.

However, it is one thing to pronounce several parts to be

one whole; it is another actually to achieve it. Although each battalion had established itself as a living, recognizable entity, it is doubtful whether at this stage there was much sense of allegiance to, and spirit of fellowship towards, the other battalions, and to the unity of the Regiment as a whole. There was indeed much to militate against this. Each battalion wore its own collar badges and bore its parenthetical additional title; the customs and traditions of the former regiments were vested solely in the battalions which each had formed; the officers and NCOs of each battalion were still largely from its former regiments; and such references as were made in conversation about other battalions tended perhaps to be caustic rather than comradely. Not that there was anything new in this; as we have seen earlier, such a spirit often existed between the regular battalions of the old county regiments from 1881 to 1947. But in an age in which the sword of defence economies was pointed almost permanently at the Army's throat, such egotistical behaviour was not conducive to survival—or so it seemed at the time.

The lead given to the Infantry by the Royal Anglians and the Royal Green Jackets, in pursuance of the Army Council's wish for large regiments, had been followed, on 31 December, 1966, by the Home Counties Brigade, which turned itself into the Queen's Regiment. Thereafter, however, no other brigade of infantry appeared eager to follow their example. But the Army Board, as the Army Council had now become, was about to exchange the light switch of encouragement for the big stick of coercion.

On 8 May, 1967, the Chief of the General Staff (whose august title had been shorn of its Imperial past) explained to all Colonels of Regiments the Army Board's new plans for absorbing, 'with the least possible pain', any probable difficulties which the future might hold in store for the Infantry, following reductions in the country's defence commitments. He was unable to say, at that date, what the extent of these difficulties might be, but he warned that they could be considerable. The Colonels were further told, doubtless to the dismay of the three who had loyally followed the wishes of Whitehall, that, in the light of the possible scale of the reductions facing the Infantry, the policy of evolution to large regiments was no longer correct. Since one of the principles behind the large regiment concept had been to produce an infantry arm which would be able to meet any change in the organizational requirements of the foreseeable future, it might have crossed the minds of the Colonels of the three large regiments, that the crystal balls available to Whitehall five years earlier had been unduly opaque.

Rather than the dreaded Corps of Infantry—a solution favoured by some military thinkers, but seldom by infantrymen, in which all battalions would be numbered consecutively throughout the infantry arm without any regimental subdivisions—the new plan demanded still larger groupings of battalions, with free movement of individuals between each group. Only by this means could reductions be absorbed, manning requirements of battalions met, and economies effected in the training of recruits. The new groups were to be called Divisions and would comprise two or three brigades/large regiments of from seven to twelve battalions. The Royal Anglian together with the Queen's Regiment and the Fusilier Brigade were to form the Queen's Division, of which more details will appear later.

This, on the face of it, seemed not too bad. The Regiment, for which so much had been done over the previous three years to make it a flourishing concern, at least was to survive, though its members might find themselves converted at any time to Queensmen or Fusiliers. Even when the first of the promised difficulties, the reduction of eight infantry battalions, was announced in July, 1967, the Royal Anglians were able to breathe again, and perhaps even gained some sardonic satisfaction at seeing three other brigades, Fusilier, Light Infantry and North Irish, hurrying to convert themselves into large regiments, the better to stomach the loss of their junior regiments or rather battalions, as these would become after conversion. So amid a sea of conversions, further amalgamations, and in two cases, the total submergence of two old regiments, the Royal Anglian Regiment steered its difficult course for another year.

Then, in July, 1968, came a new Defence White Paper, with these words about further reductions: 'The Royal Anglian Regiment. The fourth battalion. These Regiments (Queen's and Royal Anglian) will each reorganize to form three battalions'. The wording is important, if the original concept behind the decision to form the Regiment is to have meaning. What was to go was not '4th Battalion The Royal Anglian Regiment' as such, but the fourth, or junior, battalion of that Regiment, which itself was to 'reorganize' at a lower strength. To the members of the junior battalion, such a distinction may have seemed at the time no more than a semantic exercise, particularly since only four years before they had been in a well-recruited regiment with its own, old, county identity. On the other hand, had they not elected, in 1964, to become a battalion of the Royal Anglian Regiment, they might, in 1968, have been sunk without trace, like the Cameronians (Scottish Rifles) or the York and Lancaster Regiment. As it was, the concept the Royal Anglians had espoused meant that the Regiment was merely contracting and not losing a limb. A former Commanding Officer of the 4th Battalion, Brigadier Cowan, put it thus:

> 'An Infantryman has got to believe in something and for most of us that something is the Regimental Family. Our move to the large Regiment in 1964 was designed, among other things, to ensure that this Family would be able to survive the sort of problems with which we were faced in 1968. If one could not believe that the Regimental Family of the future was going to provide for the full assimilation of the 4th Battalion (including its City, County and other "supporting elements"), then it was not too strong to say that our concept of the large Regiment had been based on a hoax.'

In some ways it could be said that the prospective loss of a battalion, sad though it was, strengthened the unity of the Regiment as a whole. On 1 July, 1968, the Colonel of the Regiment, Lieutenant-General Sir Richard Goodwin, had announced in a Special Order of the Day:

> 'We now discontinue the use of subsidiary Battalion titles which denoted affiliation to a former Regiment or County. . . . As from today each Battalion becomes a component part of the Royal Anglian Regiment and takes to itself responsibility for safeguarding the former traditions and customs of all the original Regiments which formed the Royal Anglian Regiment. . . . All Battalions of the Regiment should henceforth be regarded as belonging to all eleven counties which form our Regimental area.'

To underline this progression, it was also decided that the different battalion collar badges* would be replaced by a regimental pattern of the same design as the cap badge, and that the buttons, hitherto bearing the Star and Castle, would in future carry the Royal Tiger and laurel wreath, so that the former insignia of the Royal Leicesters might be perpetuated. This final unification of the Regiment was celebrated by a Beating of Retreat by the Massed Bands and Drums at Bury St Edmunds on 13 July; the ceremony was attended by HM Queen Elizabeth the Queen Mother who was presented with a regimental brooch to mark the occasion.

The reduction to three battalions was due to be completed by March, 1971, but salvation—of a sort—came with the formation of a Conservative Government in 1970. In August of that year regiments and battalions, which had been poised on the brink of extinction, were offered the prospect of continued life, but at company strength only. This opportunity was seized upon and thus it came about that, on 9 September, 1970, the 4th Battalion contracted to form Tiger Company, an independent company within the Regiment.

* 1st Bn: Britannia superimposed on the Castle. 2nd Bn: the Sphinx, above a scroll inscribed 'Talavera'. 3rd Bn: the Eagle within the Garter. 4th Bn: the Royal Tiger within a laurel wreath. (*See* page 8.)

At the meeting held on 8 May, 1967, to explain the new Divisional plan, the Deputy Chief of the General Staff said of it: 'The aim has been to establish a structure for the Infantry which will last'. In the pious hope that this may be so, and so that posterity may judge its success, some explanation of how it was to work is necessary.

The Queen's Division was to consist of a Divisional Headquarters, the Queen's Regiment, the Royal Regiment of Fusiliers, the Royal Anglian Regiment, and a Divisional Depot. Regiments were to retain their own identities, insignia, and territorial affiliations. Officers would be gazetted and soldiers enlisted into the Division, although both would be posted initially to the regiment of their choice, whenever possible. All officers and soldiers were to be on common rolls, and could find themselves serving in any battalion, Queen's, Fusilier or Royal Anglian, of the Division. Finally all recruits would be centrally trained at the Divisional Depot.

The existing Regimental Headquarters were to be replaced by the Divisional Headquarters, presided over by a Colonel Commandant. Since he would be a senior serving officer, who would undertake this responsibility in addition to those of his normal Army appointment, he was given a full-time Deputy, a serving Brigadier, whose duties were akin to those of the former Regimental Colonels. The task of the Colonel Commandant was to direct the management of his Division on behalf of the Army Board, but he was to concern himself with major policy matters only. The detailed running of the Division became the province of his Deputy, or Divisional Brigadier as he came to be known: executive action for the administration of officers and soldiers, the planning and co-ordination of the recruiting effort and the control of all matters affecting the general efficiency of the Division as a whole.

An indication of the loss of autonomy to regiments was the new status of Colonels of Regiments and their relationship to the Colonel Commandant. The former were no longer to have any position in the normal chain of command and had to address all enquiries to the Colonel Commandant. Indeed, they were deprived of the right of direct access to the Ministry of Defence, unless so empowered by the Colonel Commandant, who would however look to them for advice on matters affecting their individual regiments. The guardianship of regimental traditions would remain the province of Colonels of Regiments, but, and this was emphasized, within the framework of Divisional policy.

The first position of power in the new hierarchy to fall to a Royal Anglian was that of Divisional Brigadier, to which Brigadier Holme was appointed on 1 January, 1968. Major-General Forrester, late Queen's, became the first Colonel Commandant on 1 July of the same year, and under his guidance the new organization took shape. On 21 September, 1971, he handed over to Major-General Dye, who had commanded the 1st Battalion in Aden.

Divisional Headquarters was initially at Colchester, but on 15 July, 1970 it moved to Bassingbourn Barracks, near Royston, Hertfordshire. Bassingbourn had been an old RAF airfield, but had been taken over in August, 1969, on behalf of the Queen's Division by Major Worthy, the GSO 2, who had been very largely responsible for ensuring that the first of the new Depots to be established was that of the Queen's Division.

The last passing-out parade of recruits to be trained at the Royal Anglian Depot at Bury St Edmunds was held on 10 October, 1969, in the presence of the Mayor, Alderman Davies, who was accompanied by the Colonel of the Regiment. Bury St Edmunds had for long been the home of the Suffolk Regiment and subsequently of the Royal Anglian. Now an era was ending and as General Goodwin, himself a Suffolk, pointed out, only the Museum, the Regimental Chapel and the Cottage and War Memorial Homes of the old Suffolk Regiment remained as a link with the past. Apart from these, and the Regimental Headquarters of the Royal Anglian Regiment, which was to remain in reduced form, the long association of civil and military was over and no longer would soldiers be a common sight in the streets of the town. Colonel Badger, who had taken over from Colonel Denny as Regimental Colonel, had the task of reorganizing the new Regimental Headquarters on a civilian basis and preparing the ground for the Divisional Depot.

The Divisional Depot began to function at Bassingbourn from 1 January, 1970, but the purpose behind the Divisional concept had been in full sway since 1968, as was witnessed, for example, in June, 1969, when the 4th Battalion, then under orders for Bahrain, received a draft of seventy-two from the Royal Regiment of Fusiliers to bring the battalion up to strength. Lieutenant-Colonel Deller was commanding the Regimental Depot at this time and became the first Commanding Officer of the Divisional Depot. This not only required tactful handling of the two other regiments but also a considerable resolve to make it work.

A writer in *Castle*, speaking of the new Depot, noted: 'We are building up a tremendous corporate spirit at this the first of the new Infantry Divisional Depots'. Whether it has been, or will be, possible to establish a Divisional spirit, comparable to the old regimental family feeling, is perhaps difficult to say, and judgement on this point may have to wait for several years yet. But when, on that summer's evening in Colchester, 450 drummers and bandsmen from nine battalions of the Division—in scarlet and blue, leopard- and tiger-skins, No. 1 Dress caps, helmets and Fusilier caps of sealskin—marched and counter-marched to the tunes of the old Army, it seemed not impossible to feel some sense of kinship, some pride of membership in this latest manifestation of the new Army, the Queen's Division, a sight for the Queen indeed.

CHAPTER 6

STRATEGIC RESERVE

'Hands off o' the sons of the Widow,
Hands off o' the goods in 'er shop,
For the Kings must come down an' the Emperors frown
When the Widow at Windsor says "Stop!"
(Poor beggars!—we're sent to say "Stop!")'
Rudyard Kipling, *The Widow at Windsor.*

THE 1969 DEFENCE WHITE PAPER stated: 'When most of our forces are stationed in Europe and trained for their operational rôle within NATO, we must ensure that we still have forces which can operate outside Europe as required'. The operations envisaged ranged from the defence of United Kingdom Colonies and Dependent Territories to those in support of treaty obligations, such as ANZUK in the Far East, in defence of our own national interests, or in support of the United Nations. The military forces available for such operations were either those stationed in garrisons overseas, such as we have seen in Chapter 3, or those based in England forming the Strategic Reserve, or the United Kingdom Mobile Force as it has been known since 1 April, 1972. The main Army element of this force is the 3rd Division, and between 1967 and 1971 three battalions of the Regiment served in it.

2nd Battalion, 1967–71

On returning to Felixstowe from Cyprus, in September, 1967, the 2nd Battalion found itself part of 19 Airportable

Brigade, based at Colchester. Conversion to the airportable rôle was quickly accomplished, but the first call on the Battalion's services made no demands on this accomplishment, nor indeed did it bear any relation to treaty obligations and the like. As the Battalion's correspondent to *Castle* put it: 'Well, we've had our first taste of active operations on the Strategic Reserve. It was called action in aid of the Civil Authorities, but the Keeping the Peace manuals don't give any instructions about cleaning out pig sties'. This bucolic activity was required of the Battalion as a result of a severe outbreak of foot-and-mouth disease in Shropshire. After a week or two's agricultural area cleaning in appalling December weather, the Battalion was relieved in time for Christmas.

In the New Year training for the new rôle started in earnest, 'B' and 'C' Companies flying to Malaysia in January for five weeks jungle warfare training. In June Lieutenant-Colonel Haycraft handed over to Lieutenant-Colonel Akehurst, who took the Battalion back to Cyprus in August for a month's training in areas it had left not a year before. At the end of the year, however, fresh fields beckoned. In December, 1968, the Battalion, less one rifle company, left Felixstowe for a nine-month spell of reinforcing the defences of one of Britain's few remaining colonies, one peculiarly linked to the Regiment by past history and present cap badge—Gibraltar.

The increase in strength of the Gibraltar garrison had become necessary in 1967 when the Spanish Government began its campaign for the return of the Rock to Spain. When the 2nd Battalion arrived, the military situation was quiet and, once the novelty of finding the frontier guards and observation posts had worn off, boredom became a greater hazard than the might of Spain.

One task that offered a complete change from the tedium of guards was the construction of Royal Anglian Way, a 500-yard walk with vantage points, affording magnificent views, along the edge of a natural escarpment 700 feet up the Rock. There had once been a path on the site connecting some old gun emplacements, but it had become completely overgrown; the new work involved clearing a mass of masonry, undergrowth and metalwork and the construction of paths, hand rails, walls and shelters. After six months' work, directed by Captain Pond, the path was opened and christened by HE The Governor and Commander-in-Chief, Admiral of the Fleet Sir Varyl Begg.

During the stay in Gibraltar many soldiers got the opportunity of visiting countries around the Mediterranean and elsewhere, including 182 men who went on adventure training expeditions to Morocco.

The Band had not been allowed to accompany the Battalion, but in order to break the musical monopoly enjoyed by the Band and Pipes of 2nd Battalion Royal Irish Rangers, the garrison battalion, a deal was done with the British India Steamship Line, which owned the old troopship *Nevasa* which was used for schoolchildren's educational cruises in the Mediterranean. In return for free band concerts on board, it was agreed that the Band would be given passage to Gibraltar. As it turned out, the Battalion got the best of the deal, for the voyage across the Bay of Biscay was so rough that the concerts given by the few bandsmen unaffected by sea-sickness were attended by even fewer children.

Another memento of the Battalion's tour in Gibraltar was a stamp, issued by the Gibraltar Post Office, showing a Royal Anglian soldier, Private Weston, in the special No. 2 Dress worn on the ceremonial frontier guard. The tour also cemented the Regiment's alliance with the Gibraltar Regiment and, since 1968, exchanges of officers and men between the two regiments have taken place.

The Battalion returned to Felixstowe in August, 1969. While it had been away, 9 Platoon, 'C' Company, which had remained behind, won the Grierson Shield, a competition for all infantry platoons in Southern Command. This competition, for which the Platoon was trained by Company Sergeant-Major Knight, incorporated the firing of all platoon weapons, a night ambush, and tests of fitness and stamina.

In November, 1969 the Battalion moved from Felixstowe to Colchester; it was the last infantry battalion to serve in Felixstowe and indeed the last Regular Army unit to be stationed in Suffolk. It had hardly settled into Hyderabad Barracks when it was away again; this time to Kenya for a month's preparatory training prior to a major exercise later in the year. Many men found time to climb Point Lenana (16,355 feet) and a party under Captains Gowing and Pond visited Lake Alice high on Mount Kenya. This lake had been so named in 1938 to mark HRH The Duchess of Gloucester's marriage, and a plaque was left by the 2nd Battalion party to commemorate their visit thirty-two years later. After returning to Colchester in February, 1970, a letter was received from the British High Commissioner in Kenya; this singled out the Battalion's performance there for special praise and concluded: 'When a British battalion conducts itself in this fashion in Kenya it does immense good to the British image'.

Preparations immediately began for Exercise '*Bersatu Padu*' in Malaysia. This, the largest exercise ever to be held in the area, had been designed to show the British ability to reinforce the theatre quickly, and, once there, to work in unison with other SEATO forces, despite the British withdrawal from the Far East. The whole of 19 Brigade was involved in the exercise, together with troops from Australia, New Zealand, Malaysia and Singapore. The Battalion flew first to Singapore, then moved up to Malaysia for two months' jungle training, followed by a two-week international exercise; this started with a counter-insurgency phase, building up to limited war. In a report on the exercise, *The Economist* referred to the Royal Anglians as 'that excellent battalion'.

Soon after returning to England in September, 1970, Lieutenant-Colonel Akehurst handed over to Lieutenant-Colonel Gerrard-Wright. The latter, having been told that an emergency tour in Ulster (where the military involvement had started in 1969) was unlikely in view of the Battalion's forthcoming move to BAOR in the spring of 1971, found himself, in the event, leading the Battalion into its first experience of the Irish problem, an experience which has since become all too familiar, and which is covered in Chapter 9.

3rd Battalion, 1967–70

It will be recalled that the 3rd Battalion had served briefly in the Strategic Reserve in August, 1966, before its emergency tour in Aden from October, 1966–May, 1967. On returning, it became part of 5 Airportable Brigade at Tidworth. The intensive training demanded of Strategic Reserve battalions soon got under way: exercises in the United Kingdom training areas of Otterburn, Stamford and Sennybridge; overseas to Kenya in early 1968, and, breaking fresh ground, to France in August, 1968, to Denmark and Germany in the winter, and to Iceland in the spring of 1969. All these manoeuvres gave the Pompadours a wide variety of terrain and climate in which to practise the military arts.

A change and a chance to perform a more ancient skill occurred in October, 1968, when the Battalion paraded to Troop the Colour before the Colonel of the Regiment. This parade marked the tenth anniversary of the formation of the 3rd East Anglian Regiment and, in a series of tableaux arranged to portray the Battalion's rôles over the period, it was interesting to observe how almost all the Infantry's equipment had changed since 1958. Sadly, the Deputy Colonel-in-Chief, HRH The Princess Margaret, was prevented by illness from attending this day, but she visited the Battalion in June the following year.

In December, 1968, the Battalion moved to Normandy Barracks, Aldershot, to join 16 Parachute Brigade, not, however, with parachutes, and the following June Lieutenant-Colonel Dymoke handed over to Lieutenant-Colonel Burch. The Battalion now had the task of conducting trials on new tactical concepts of deep penetration tactics in Europe. This led on to the testing of certain theories on the use of infiltration tactics in the night attack, which had been developed by the Brigade Commander, Brigadier Eberhardie, himself a former Royal Norfolk. This training, which required a high degree of skill at navigation and night movement, culminated in a Brigade exercise on Salisbury Plain in September, 1969, which, despite some navigational dramas and very heavy rain, was rated a great success by the Brigadier.

Shortly before this exercise the Battalion had another taste of ceremonial by providing street lining parties in Victoria Street and Parliament Square for the State Visit of President Kekkonen of Finland, whose own troops the Pompadours would shortly be meeting in very different surroundings.

The 3rd Battalion was now about to undertake an entirely novel rôle for the Regiment: that of the British battalion in the United Nations Force in Cyprus (UNFICYP). The tour of duty was to be for six months and, to signify their new status as international troops, the pale blue beret with United Nations badge was adopted instead of the normal Royal Anglian headdress. This unique task has been described in an account written by the Commanding Officer, Lieutenant-Colonel Burch:

> During this time the Batttalion was fully under United Nations command: it took its orders from the Force Commander— an Indian General—and operated on policy and procedures laid down by the United Nations Secretary-General in New York. This task of peacekeeping was quite unlike anything the Battalion had done before and very different from the more familiar counter-insurgency role the Battalion had mastered in Aden some three years earlier.
>
> The United Nations Force was in Cyprus to maintain peaceful conditions between the majority Greek Cypriot population, who provided the Cyprus Government, and the much smaller Turkish Cypriot community. Since the violent intercommunal fighting during the winter 1963–64 the latter had left all their

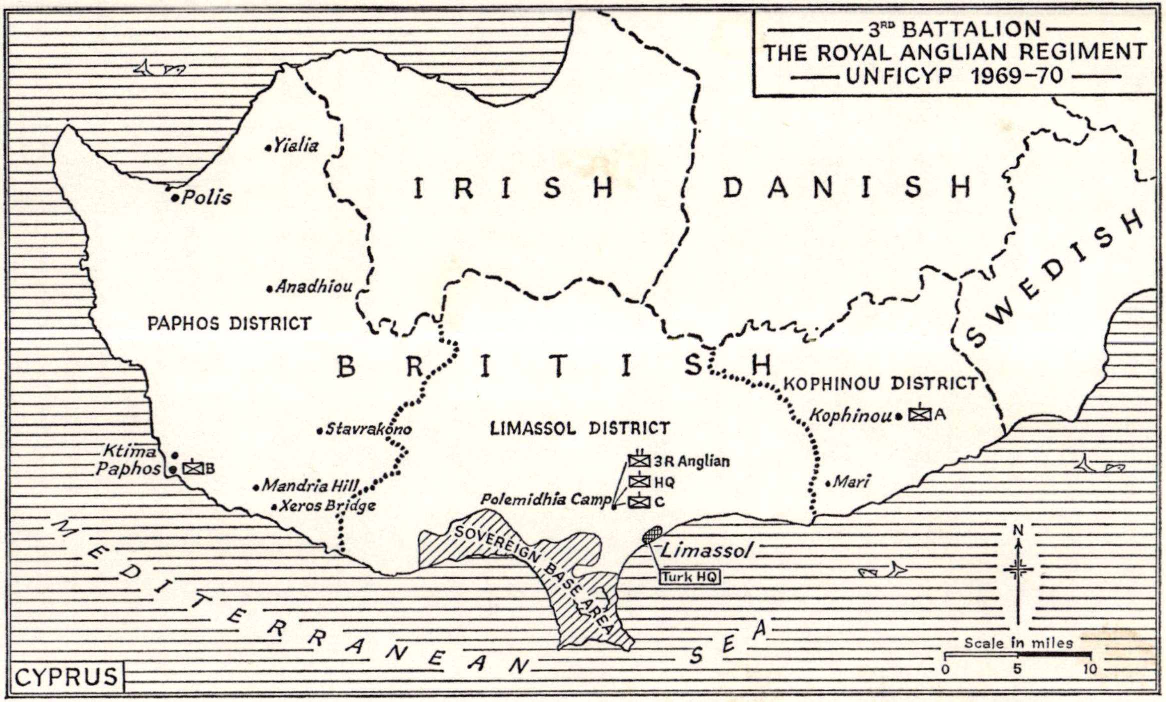
3RD BATTALION
THE ROYAL ANGLIAN REGIMENT
UNFICYP 1969–70
IRISH
DANISH
SWEDISH
BRITISH
PAPHOS DISTRICT
LIMASSOL DISTRICT
KOPHINOU DISTRICT
Yialia
Polis
Anadhiou
Stavrakono
Ktima
Paphos
B
Mandria Hill
Xeros Bridge
3R Anglian
HQ
C
Polemidhia Camp
Kophinou
A
Mari
SOVEREIGN BASE AREA
Limassol
Turk HQ
MEDITERRANEAN SEA
N
Scale in miles
0
5
10
CYPRUS

official government posts and set up their own illegal administration and defence force, known as the 'Fighters'. They ceased to recognize the Government and sealed themselves off in separate, protected, self-administered enclaves while lengthy negotiations to resolve their differences with the Greek Cypriots continued over several years in the peace maintained by the United Nations. When the Battalion arrived the situation on the surface, seemed quiet enough. However, it was soon apparent that a real depth of hatred existed between the two communities which maintained a close armed watch on each other ready to take the slightest advantage, and prepared, if they could, to outwit the United Nations peacekeepers if this would help them to score over the other side. A *modus vivendi* existed between the two sides based on unwritten local understandings which the peacekeepers had to learn and ensure were maintained.

The United Nations Command was divided into a number of geographic areas of responsibility and each national army contingent (Canadian, Danish, Irish, Finnish, Swedish and British) was allotted a zone in which it was supported by a United Nations Civil Police contingent. The Battalion's zone was 1,500 square miles in area and included the important town of Limassol containing some 10,000 British dependents from the Sovereign Base Area where, by coincidence, the GOC was Major-General M. W. Holme, our Deputy-Colonel. The Battalion's zone was divided into a number of rifle company areas of responsibility called Districts with one headquarters at Kophinou (Major Gill—'A' Company), a sensitive area where twenty-two Turkish Cypriots had been killed two years earlier in an incident which had nearly brought about the invasion of Cyprus by Turkey. At Polemidhia Camp, Limassol, Major Gee ('C' Company), who did such sterling work after part of Limassol was devastated by a tornado at Christmas, was primarily responsible for peacekeeping in the town, where the substantial, fortified Turkish enclave was a constant source of friction and potential danger spot. This required a small platoon garrison permanently stationed in a house in the area where the two communities adjoined most dangerously. The third District Headquarters was commanded by Major Ford ('B' Company) and located at Paphos. This covered a rural wine-growing area with many scattered Turkish Cypriot communities; the need to watch these demanded a more permanent dispersion than the other districts. The Battalion had under command fifty Australian Civil Police located in detachments throughout the zone whose job was to liaise with civil police elements of both communities and to supplement our patrols. The Australians were tough, competent men, most of whom seemed to come from Sydney. They were great fun and although their idea of being under command was quite different from what is taught at the Staff College a very close and happy relationship existed with them. Their assistance and advice was invaluable on many occasions.

The Battalion's job was to keep the peace between Greek and Turkish Cypriots in its 1,500 square miles and to create an atmosphere between them conducive to the conduct of political negotiations at Government level. This meant preventing incidents, or, by quick action stopping the escalation of any which did occur; to do this meant knowing what was going on, anticipating what might happen next, and being able to pass information quickly and accurately. Methods of operation were based on a network of permanent Observation Posts, usually of section strength, and Battalion and Australian Police patrols throughout the Zone, covering sensitive spots and areas of permanent intercommunal confrontation. The task of OPs and patrols was to watch and report, particularly that both sides were observing local agreements. If they could, they were to settle local problems arising on such matters, for example, as freedom of movement by one side through the other's area, or complaints about improvement in the opposition's fortifications or excessive training. Sections spent three weeks in each OP rather than remain in the same place for six months as some other nations did, and in this way the dangers of boredom were avoided; nonetheless the soldiers soon came to know the locals in each area well, and spent many hours drinking coffee with them listening to the latest grievances and gossip. Relations were usually very friendly even to the point that on several occasions off duty soldiers were invited to weddings. The OP 'arms plot' became quite complicated and movement between most posts was by helicopter; rotation day every three weeks was an involved process which was organized without mishap by the 2IC, Major Randall. Finally most OPs had a dog, regularly vetted by the Regimental Medical Officer, as a permanent pet and the soldiers became very attached to these. Many dogs had been at their OP so long that it was said that they knew more about the business than the Section Commanders: the infamous Blackie at Mari OP distinguished himself by biting the Commanding Officer.

The next method of operation was by the constant maintenance of close liaison with leaders of both sides at all levels from Section Commanders talking to the leaders in the village where their OP was located to the Commanding Officer dealing with Government District Officers and the equivalent Turkish Cypriot leaders. In the event of an urgent local dispute the aim was to try to find a solution acceptable to both sides and to settle the matter before it could escalate to a shooting incident. This involved considerable negotiations with each side separately, many visits between them, many promises broken by both sides and much distortion of the facts. A great deal of patience was needed over these problems, many of which seemed quite trivial to the neutral observer, but which could, and frequently did, extend over many months, influenced by pressures from the respective Greek and Turkish leaderships in Nicosia and in an atmosphere heavy with the threat of violence and shooting. It was often necessary to establish temporary OPs at trouble spots. The bulk of negotiating work was, however, conducted by the Company Commanders at routine meetings held, quite separately of course, with both sides, at which inter-communal complaints and United Nations liaison matters were raised.

The final method of operation was the maintenance of reserves in different locations and at varying degrees of notice so that new OPs could be quickly established as required, or that other areas, including another nation's zone, could be reinforced; on one occasion the Battalion had to send reinforcements to the Canadian battalion in Nicosia. In the event of real trouble these reserves made it possible to interpose troops between both sides as a deterrent and as an effective reporting agency. There was, however, no question of ever using force ourselves except in self-defence, though patrols, sentries, OPs and reserves all carried arms.

The dispersed state of the Battalion and the frequent rotation of sections and platoons made administration difficult. To ease the problems company headquarters were left in the same location for six months and given an administrative increment to make the company self-supporting; platoons were rotated between them.

'A' Company headquarters, located in a small, overcrowded, wired-in, dusty compound beside the main Nicosia–Larnaca road junction, came off worst, but took such pride in making the best of their discomfort that they created the 'Kophinou Compound' tie with a motif of barbed wire coils. Anyone serving in the compound for a hundred consecutive days became eligible to buy the tie, much to the benefit of the company fund. 'B' Company with its splendid villas at Paphos did rather better.

The wide dispersal of administrative elements added to the responsibilities of Headquarters Company officers at Polemidhia (though their critics claimed they were lessened!) and a regular helicopter marathon run was necessary each week for the Paymaster (Major Lunn), Medical Officer (Captain Melling, RAMC) and Padre (The Reverend Paul Mears) to deliver the pay to each camp and OP, to inspect its hygiene and to fortify the souls of the occupants, though not necessarily in that order of importance. Keeping transport serviceable on the rough roads was a difficulty cheerfully and efficiently overcome by the MTO (Captain Franks) while the Quartermaster (Captain Denny) coped with a huge miscellany of problems, including the provision of United Nations Medals which, in this case, literally did come up with the rations. Finally, the WRVS provided Kathy Clarke who also did 'the OP run' and whose cheerful work was greatly appreciated by the Battalion.

Apart from what might be called its technical duties the Battalion was concerned with ensuring good social relations with Greek and Turkish Cypriots and with the other contingents of the United Nations Force; despite language difficulties this was largely achieved through social events and sport. Entertaining Greek and Turkish Cypriots was not without protocol dangers as they would not attend social events together unless the events were so large they could avoid meeting each other. Unless treated exactly the same as his opponent (who, in any event, he would not acknowledge) a local leader was liable to take offence and walk out of a social function. Despite all these hazards, however, the Battalion successfully beat retreat by searchlight and held a band concert in the Nicosia Municipal Theatre for UNFICYP, attended by the British High Commissioner and Force Commander, and many public band engagements for both the local communities and other United Nations contingents. A parade was held at which UNFICYP peacekeeping medals, awarded to all soldiers completing three months with the Force, were presented to a representative element of the Battalion by the Force Commander, Major-General D. Prem Chand of India. The Officers' Mess received twenty-five visits from officers of two-star rank and above, including the British and Canadian High Commissioners, the Force Commander and his predecessor General Martola of Finland (under whom the Battalion also served for three months), the Commander British Forces Near East, our Deputy Colonel, Major-General Holme and the Director of Military Operations from the Ministry of Defence. All this was admirably handled by the mess staff and in particular by Corporal Ruffle, the cook. '*Quel Chef magnifique*', the French PR chief called him, to which Corporal Ruffle replied that in all his service no one had ever called him a chef!

Despite all the difficulties of dispersion the Battalion played a very full part in the sporting life of UNFICYP and the British Sovereign Base Area, including taking part in the gruelling Finnish military obstacle race competition. Perhaps the high point in the sporting calendar was the UNFICYP falling plate shooting competition, organized by the Battalion, in which its team reached the final shoot against its great friends and rivals the Irish Battalion only to achieve the almost impossible result of a tie. In the shoot off the guests diplomatically won by a split second.

So a very full and interesting tour came to an end. The peace had been kept for six months, with only a few incidents important enough to be reported to New York, thanks to the good humour, tolerance and tact of the soldiers and the patience of the officers conducting negotiations. Lives were all the richer for the many friendships made among fellow peacekeepers from other nations and among the two sides between whom the Battalion had kept the peace for six memorable months.

At the conclusion of the Battalion's tour, the UNFICYP Force Commander, Major-General Prem Chand, wrote to Colonel Burch thanking the Battalion for its good work and continuing:

> 'May I also take this opportunity of complimenting the Battalion on the very high standard achieved in all aspects of your duties, and especially on the excellent relations maintained at all levels with the good people of Cyprus.
>
> 'More personally, as I said at your splendid Medal Parade earlier this month, to have had an opportunity of serving with British Troops after a period of over thirty years has been a unique and treasured experience for me, and I shall always look back upon our association in Cyprus with the happiest of memories.'

While the Battalion had been in Cyprus, it had been warned for service in BAOR as a mechanized battalion in December, 1970. Thus, after returning to Aldershot in April, 1970, an entirely new and different task had to be undertaken: conversion to the mechanized rôle and the massive programme to produce the drivers and operators required. Conversion was not made any easier by having only half the complement of Armoured Personnel Carriers (APCs) and by the shortage of tracked training areas around Aldershot. Driving APCs around the Hampshire roads was not without hazard, as Colonel Burch recalls:

> 'An APC (failing to carry its vertical exhaust) with its chest-high horizontal side exhaust on the near side of the road (unlike BAOR where it would be on the other side away from pedestrians) overtook an attractive young lady in Crookham. Her nylon dress was showered by the hot exhaust fumes and almost disintegrated. The CO, on receiving an irate complaint and demand for compensation, despatched a handsome young Captain (Norbury) with a blank cheque drawn on the CO's fund. The young lady ended up with an expensive replacement dress and was very happy!'

The mechanized training culminated in a week's manoeuvres on Salisbury Plain with a squadron of the Royal Hussars. This exercise was sadly marred by the tragic death of Major Gee, an extremely popular and able officer.

Before leaving for Germany, the Battalion provided the British Army team for the 1970 NISHAN (CENTO skill-at-arms) competition against teams from Iran, Turkey and the United States. Under the leadership of Major Cridland, and helped by a fine winning shoot by the General Purpose Machine Gun pair, Lance-Corporal Sykes and Private Elliot, the team achieved second place with 3,150 points against Iran's 3,268; this was the best result hitherto scored by a British team in the competition.

One stalwart of the NISHAN team was Corporal Aves, a member of the Army Shooting VIII, who had reached the Army Hundred at Bisley in 1964 and 1969, and in 1970 came third in the Army Championships, winning the Bronze Jewel.

Success in a different sphere was gained by Private Anthony, who was second in the 200 metres against the German Army in 1969 and also in the British Army Championships at Aldershot in 1969 and 1970 with a consistent time of 22·0 seconds.

4th Battalion, 1968–70

On returning home from Malta in July, 1968, the 4th Battalion went to Chatham as part of 2 Infantry Brigade, whose headquarters were at Shorncliffe. The year the Battalion was to spend at Chatham was overshadowed by the knowledge that its life was limited, as described in Chapter 5. Nevertheless the familiar training cycle went ahead with exercises in the United Kingdom training areas and, in the case of Support Company, as far afield as Malaysia. In September Lieutenant-Colonel Carter handed over to Lieutenant-Colonel Holloway, one of whose first tasks proved to be a highlight of this period in England.

On 12 October, 1968, the 4th Battalion marched through the City of Leicester, exercising for the first time its right to do so since the granting of the Freedom of the City to the Battalion in 1966. The march and the address by the Lord Mayor, Alderman Bower, were marred by pouring rain but nevertheless the streets were lined by hundreds of Leicester citizens. The salute was taken by the Lord Mayor, who was accompanied on the dais by the Deputy Colonel-in-Chief, HRH The Duchess of Gloucester, the Lord Lieutenant, the Colonel and Deputy-Colonel of the Regiment. After the parade, the Battalion was entertained to a civic lunch, followed by a First Division Football Match between Leicester City and West Bromwich Albion.

Five days before this parade, the Battalion had been warned for a nine-month tour in Bahrain starting in August, 1969. It has already been observed in the last chapter how the Battalion was reinforced for this move by a draft of Fusiliers. Another indication of how the reinforcement system was working was the influx of men from East Anglia into what had hitherto been predominantly a Leicestershire battalion. A humorist in *Castle* saw fit to acknowledge this fact with the publication of an East Anglian phrase book, 'to assist those who were unfortunate enough to have been born west of the Ouse or south of the Deben'. It included such 'Dos and Don'ts' as 'Don't be discouraged if you don't at first understand a single word' and 'Simple Phrases' like: 'That's strange, isn't it? — Thessa rummun hent ut?'

Armed with such useful hints, the 4th Battalion arrived at Hamala Camp, Bahrain, on 13 August, 1969, taking over from 1st Battalion The Cheshire Regiment.

After the British departure from Aden, and the discovery of more oil in the Persian Gulf, British forces in the latter area had been increased. At the time of the 4th Battalion's arrival in Bahrain, there was another battalion and a battery in Sharjah, plus supporting elements, all commanded by HQ Land Forces Gulf, which itself was in Bahrain.

Since the island of Bahrain was a sovereign state, the presence of a British battalion was somewhat anomalous. To avoid alarming the sensibilities of the local population, its existence and activities had to be kept fairly subdued. The Battalion had a rather nebulous internal security rôle, but political considerations, added to the small size of the island, imposed considerable limitations on training.

Fortunately there was greater scope on the mainland, where exercises could be held in the Trucial States and in Muscat and Oman. Apart from these exercises, platoons were sent away on their own into Oman to carry out a series of patrols known as Omex. These patrols were based on a company of the Sultan of Muscat's Armed Forces (SAF) at Saiq in the Jebel Akhdar and gave platoons a change of air from the heat of Bahrain. Each platoon commander organized his own training and administration and carried out liaison with the local population and SAF, helping where necessary with the training of the latter. In April, 1970, a composite company group under Major Robinette undertook a fortnight's patrol and covered just over 1,000 miles, mainly through the mountains of northern Muscat.

The City of Leicester maintained its usual close interest in the Battalion during its stay in Bahrain, demonstrated by the visit, from 14–19 January, 1970, of the Lord Mayor, Alderman Edward Marston, who was also received by the Ruler, Sheikh Isa bin Sulman al Khalifa. The latter visited the Battalion on 15 April and presented a gold ceremonial sword to the Commanding Officer. His courtesy was returned on 12 May by the presentation, on behalf of the officers, of a silver tiger. These exchanges marked the end of the Battalion's tour; ten days later the return to England began.

In addition to the 4th Battalion's service in the Persian Gulf, a number of individual officers of the Regiment have made a significant contribution to the stability of the area while serving with SAF. An account of their activities appears in Appendix V. For security reasons, this account is of necessity brief.

The Battalion re-assembled for work in Chatham on 6 July and the sad process of disbandment began almost immediately. Then, on 27 August, came the reprieve for at least part

Led by the Band and Drums of the Junior Soldiers of the Regiment, ie final passing out parade of recruits marches under Abbey Gate, ury St Edmunds.

7. Soldiers dressed in 'uniforms through the ages' hold the ground at Blenheim Camp on 13 July, 1968. The Bands and Drums of all four Battalions Beat Retreat in the presence of the Colonel-in-Chief.

8. The Governor of Gibraltar opens the reconstructed path now called 'Royal Anglian Way'.

A mechanized parade in Munster.

). Volunteers of the 5th Battalion on training for the Rhine Army
serve role.

11. 2nd Battalion soldiers practise river crossing in the Malaysian jungle, 1968.

12. 1st Battalion soldiers bridge-building in the rain forests of Gathuiru on the foothills of Mt Kenya, 1973.

of the Battalion—the announcement that an independent company might be retained. On 9 September Tiger Company was formed.

Shortly before this, the old 17th Foot enjoyed a brief resurrection, as if to bid farewell to its descendants. At the 21st Edinburgh Tattoo, Lieutenant Towns and twelve soldiers of the 4th Battalion appeared dressed in the uniform of 1714, to re-enact the ceremony of the installation of the Governor of Edinburgh Castle at a time, during the reign of Queen Anne, when the garrison was found by Wightman's Regiment (17th Foot).

Tiger Company, 1970–73

Tiger Company on formation consisted of 105 all ranks, organized into a Headquarters, an Administrative Platoon and two Rifle Platoons. Under the command of Major Heggs, the Company was to provide the demonstration platoons for the Royal Military Academy, Sandhurst, and the Mons Officer Cadet School. The Sandhurst platoon, under Lieutenant Shalders, started work straight away, but Lieutenant Field's platoon had time for a month's work in Ulster with the 1st and 2nd Battalions before starting at Mons in April, 1971.

In December, 1971, the Company exchanged this task for a six-month tour in Gibraltar, carrying out a similar rôle to that of the 2nd Battalion in 1969. For this the Company increased its versatility by forming a Corps of Drums and a Support Platoon of two 81 mm mortars, two Wombat anti-tank guns and three General Purpose Machine Guns. Many long hours were spent in Observation Posts watching the Spanish frontier, and the tedium evoked by these duties can be gauged by this entry on one sentry's log sheet:

> 'Serial 92. 15.30 hrs. Major Heggs visited OP to give me my pay. He fell over going down the steps. Sgt Kingshott found it funny, I don't think the Major did.'

In addition to these tasks, the Company received many visitors, including HRH The Duchess of Gloucester and the Lord Mayor of Leicester, Alderman Percy Watts; performed the Ceremony of the Keys; refurbished Royal Anglian Way and undertook such training as the limited facilities of Gibraltar would allow.

On returning to England the Company found itself with a new rôle but reduced almost to cadre strength, consisting only of the officers, NCOs and six privates, the remainder being posted to the other battalions. Its new job, which began in September, 1972, was the training of junior soldiers at the Junior Infantrymen's Wing at Canterbury. Though it was not without interest, this undertaking, and more especially the attenuated strength of the Tigers, must have raised questions in the minds of many as to the future existence of the Company as a separate entity. However, all was not lost and, by November, 1973, the Company had re-formed under the command of Major Swallow as the only independent infantry company in the Army, with the prospect of an operational rôle in 2 Infantry Brigade.

CHAPTER 7

MECHANIZED INFANTRY

'A force of selected infantry, available either to support our own cavalry or sufficiently mobile to act alone in the face of an enemy's cavalry.'
'Mounted Infantry'—
Naval and Military Magazine, 1884.

OVER THE CENTURIES infantry have always marched in the dirt thrown up by cavalry, whether the latter rode superciliously past on their horses or drove clamorously by in tanks or armoured cars. Though the British Infantry can probably reflect that more battles have been won by their stoicism and endurance than have ever been gained by the dash of the cavalry, nevertheless the latter has always been able to maintain an aura of superiority over the man on foot, even though this has often been based on social and sartorial advantages rather than on purely military grounds. However, in an increasingly democratic and professional service, such advantages have been for some time of diminishing importance, and modern warfare has ensured that, in more than one sense, the infantryman is now on level terms with the mounted man. The advent of tactical nuclear weapons on the battlefield has postulated the mechanization of the infantry, and the introduction of the Armoured Personnel Carrier has elevated the foot soldier out of the dust thrown by galloping hooves or the muddy ruts of the tank tracks.

The modern land battle is fought by mixed battle groups of all arms, the composition of each being flexible and dictated by the task immediately to hand; where armour predominates, the battle group will have an armoured commander and take its style from the armoured regiment and vice versa. Thus, in the future, a company of a Royal Anglian battalion might save the day on some stricken field fighting not, as in the past, as part of *N*th Battalion The Royal Anglian Regiment, but of, for example, the 16th/5th Lancers Battle Group; as the fighting moved into thicker country, the same company might find itself a few hours later as part of *N*th Royal Anglian Battle Group.

The most likely scene of such eventualities is, of course, the North German plain, where the British Army of the Rhine stands on watch, as part of the NATO alliance, to deter and, if necessary, to engage the potential enemy in the East. Since 1965 Royal Anglian battalions have played their part in this rôle, first converting themselves from garrison or airportable battalions into mechanized infantry, then meeting and training with their armoured comrades, and finally rolling out in their APCs in their mixed battle groups to engage in the countless mock battles which maintain the Rhine Army as the highly efficient part of the NATO shield that it is.

1st Battalion, 1965–70

Suitably the 1st Battalion was the first to convert to the mechanized rôle. After leaving Aden in September, 1965, the Battalion joined 7 Armoured Brigade in Celle. The implications of their new task were summed up by the Battalion's correspondent to *Castle*:

> 'The transfer from Aden Brigade to 7 Armoured Brigade provided a complete contrast in climate, living conditions and routine. It also meant that in addition to revising our infantry skills to meet the training requirements of European war in a nuclear setting, we had also to acquire technical knowledge and expertise corresponding to that of an armoured regiment. This additional knowledge had moreover to be absorbed quickly and thoroughly enough to fit us for fully integrated armoured

operations with our new Brigade by March; we could not afford to be just passengers while we carried out the conversion process at our leisure.

Conversion to the new rôle meant that a Section Corporal had to command and handle not only ten men and their personal weapons and a radio set, but also an armoured vehicle costing nearly £20,000 fully equipped, with all the technical and servicing commitments involved in "going mobile". The same increase in scope and complexity applied at each stage of the chain of command; the Technical Quartermaster now had on charge over two and a half million pounds worth of equipment. The average taxpayer would be entitled to ask if he was getting his money's worth.'

The success and speed of the Battalion's conversion were demonstrated in January, 1966, when it took the field with over eighty APCs in a four-day exercise, two months before the deadline in March by which it had to be operational. The final test of the conversion training occurred in September, when the Battalion took part in a 1st Division exercise with German, Belgian, Dutch and Danish troops. In this, its first 'battle', the 1st Royal Anglian Battle Group was joined by 'B' Squadron, 11th Hussars, and No. 1 Company of the Danish *Prinzens Livregiment*, but lost 'A' and 'B' Companies to the 11th Hussars and 2nd Royal Tank Regiment Battle Groups respectively. Later in the exercise 'A' Company rejoined the Battalion Battle Group for an assault crossing of the River Weser by night, the first time that an APC battalion had conducted such an operation.

As mechanized training is a continuous process in BAOR, some variety is required to prevent troops becoming stale. A change of scene is often beneficial and the Battalion found this in early 1967 with an exercise in Libya and later in the year a fortnight was spent in Denmark, training with the *Falsterske Fodregiment*.

To ensure that the time-honoured infantryman's skill of marching was not entirely lost, companies entered for the Nijmegen and Hoogeven marches, held annually in Holland, events which not only exercised the lower limbs, but also afforded unlimited opportunities for international fraternization of considerable variety. Survival training, skiing, escape and evasion exercises, plus the usual round of sporting activities, all gave the men of the Battalion a respite from training for mechanized warfare.

One task which perhaps more than most brought home to the soldiers the realities of life in Central Europe was patrolling the frontier with East Germany, a task which gave occasional glimpses of the potential enemy and his weapons. The soldier's first sight of the frontier came as a surprise, as a 'B' Company correspondent noted in *Castle*:

> 'Our road came to the crest of a small hillock, giving us a sudden view and there, a few hundred metres ahead and to either side, was the line of "Goon" towers. Between us and the towers was the sombre network of rusting barbed wire and, behind the wire, the sinister empty space of the minefield belt.'

These patrols, which were of platoon strength, had the task of escorting officers of the British Frontier Service, who were responsible for supervising the integrity of the border. The patrol would move down the west bank of the Elbe for eighty miles, ending on the Baltic coast at Travemunde where, in the words of *Castle*:

> 'The Baltic was a cold grey, stretching north into a damp fog, where gallant little trawlers were fighting mountainous waves, engulfed every so often by freezing spray. The view was wild and beautiful. Lashed by icy rain we surveyed the desolate East where a sign read "Attention! Danger to Life! If you pass the demarcation line the SBZ soldiers will shoot."'

In October, 1967, 7 Armoured Brigade, organized into four battle groups (11th Hussars, 2nd Royal Tanks, 1st Royal Anglian and 1st Black Watch), took part in an important exercise designed by the Operational Research Establishment to assess the value of helicopters armed with guided weapons in the anti-tank rôle. This was followed by the training in Denmark, mentioned earlier, and then, in the New Year, the Battalion suddenly changed its hunting grounds from the North German plain to the moors and bogs of Northern England.

Government economies in defence required, among other things, the withdrawal of a number of troops from Germany to stations in England, where they would continue in the mechanized rôle, returning annually to BAOR for two months' training. Thus, in February, 1968, the 1st Battalion, now under the command of Lieutenant-Colonel Emsden, who had taken over from Lieutenant-Colonel Creasey in September, 1967, left Celle and 7 Armoured Brigade for Catterick and 6 Infantry Brigade.

Life now followed the routine of an infantry battalion in England, with the difference of still being fully mechanized and the annual training in Germany each September. The North Riding of Yorkshire was soon made aware of this new type of infantry in its midst, by a demonstration designed to show the modern Army in action called 'Northumbrian View'. This was an exciting spectacle and so successful that a repeat performance in 1969 drew an audience of 55,000. The Battalion also showed itself off in the Regimental area in 1969 by giving displays at the Lincoln and East of England Shows.

After the 1969 training in Germany the Battalion returned to Catterick to prepare for an emergency tour in Northern Ireland and to greet a new Commanding Officer, Lieutenant-Colonel Jackson, who took over in November. In the event, the emergency tour did not materialize; instead the Battalion was warned for a two-year tour in Ulster starting in July, 1970. Their destination was Londonderry, where they will be met again in Chapter 9.

After nearly five years as mechanized infantry, the Battalion could reflect, with some satisfaction, on the remarks in its 'Fitness for Rôle' report for 1969, which pronounced it to be, 'one of the finest mechanized battalions in the British Army'.

3rd Battalion, 1970–73

Six months after the mechanized rôle was given up by the 1st Battalion it was assumed by the 3rd, which arrived in Germany from Aldershot in December, 1970. Its station was Paderborn, where it formed part of 20 Armoured Brigade.

One of the major problems facing the Battalion was in no way connected with mechanized training but was greatly concerned with the well-being of the soldiers. The 3rd Battalion had 370 families for which there were only forty-two married quarters, a disappointing start to life in a new station. However, thanks to the efforts of the Battalion Families Officer, Major Ingle, a sufficient quantity of other accommodation was found so that within a few months all were satisfactorily housed.

The Battalion soon settled down to its new rôle, making itself familiar with the German training areas, known so well to thousands of Rhine Army soldiers: Sennelager and Soltau, where the terrain was not dissimilar to Aldershot; Putlos and Hohne, the Meccas of support weapons; Haltern and Hameln, the latter where the crews of the APCs learnt to float their vehicles across the Weser.

Exercises were held with Danes, Germans, Americans and French, as well as with the Battalion's affiliated armoured regiment, the 9th/12th Royal Lancers. During the summer of 1971 the APCs were exchanged for helicopters to test the best method of carrying out *coup de main* operations. These tests proved their worth in the September manoeuvres when an exercise culminated in an excellent operation of this nature by Major Cridland's 'C' Company, supported by an attack by 'A' Company, with American tanks and engineers, under Major Cross. In the 4th Division exercise that followed, the Battalion, with 'B' Squadron 9th/12th Lancers, put in such a spirited counter-attack that a Danish regimental commander on the receiving end complained to the GOC that 'if this was training, what could war be like?' As Lieutenant-Colonel Burch has recorded, 'Happily the General told him!'

During their first year in BAOR, 3rd Battalion sportsmen

won the 4th Divisional Athletics and Swimming Championships, but it was left to the Battalion Football Team, captained by Corporal Allen, to score the Pompadours' most notable success by winning the BAOR Infantry Cup Final, beating 1st Battalion The Royal Regiment of Fusiliers, the holders, 2–1 after extra time.

In November, 1971, Lieutenant-Colonel Burch handed over to Lieutenant-Colonel Hall-Tipping, and the Battalion put away its APCs in preparation for an emergency tour in Ulster in April, 1972, which will be covered in Chapter 9. The Battalion returned to Paderborn in early September and, after going on leave, resumed the mechanized rôle with training at Soltau, followed by the 4th Division exercise. The rest of the year followed the familiar BAOR pattern, cadres, classification, field firing, skiing and the annual 'Fitness for Rôle' inspection. 1973 followed a similar pattern: Ulster from 27 March to 25 July, then back to the APCs for the September exercises, finishing up with a large-scale NATO manoeuvre held in Denmark in November.

During 1973 the Pompadours again won the BAOR Infantry Football Cup, going on to challenge and beat the United Kingdom champions, 1st Battalion The King's Own Border Regiment. In the boxing world, Private Maxwell won the Combined Services Light Middleweight title in April and, a month later, the ABA title. He later represented his country in a match against Hungary, coming on as a substitute to become one of the only three Britons to win in the match.

2nd Battalion, 1971–73

By mid-1971 the Regiment had two battalions in BAOR when the 2nd Battalion arrived from Colchester in May, following their first tour in Ulster, to join 4 Guards Armoured Brigade at Munster. The change from life in the Strategic Reserve was greeted with mixed feelings, as the Battalion explained in *Castle*:

'The return to Germany, with its well-established Garrisons, its long tour and its long planned exercises, seemed to us of "Strat reserve" to bring promise of a better life. "Proper soldiering," said the 1st Bn types, "with the smell of diesel in your nostrils, the clank of tracks in your ears and the firm black mud of Soltau under your feet." There have been times when we would have volunteered as a man to return to the improper soldiering of Stratco or Belfast or anywhere if only the phone would stop ringing and the paper stop piling!

' "So much to do — so little time" always seemed rather a trite saying but it has now replaced "There's no place like home" as the No. 1 favourite.'

In addition to picking up the usual BAOR routine, similar to that experienced by the 1st and 3rd Battalions, the 2nd was given a rôle which was unique among mechanized infantry: that of Nuclear Convoy Escort (NCE). This involved the provision of guards and escorts whenever British elements of the land-based deterrent were moved. This was an operational task, but has also to be performed regularly in peacetime as stocks are changed over and inspected. Initially the novelty and singularity of this rôle gave it some interest value but, as time went on, the familiarity of its performance bred, if not contempt, at least boredom.

The Battalion's first foray into the field as a mechanized battalion took place in late July with three weeks' training in the Eifel Mountains with a squadron of the Royal Danish Life Guards, an attachment in keeping with the Battalion's affiliation with the British Household troops. The elevated company in which the Battalion now moved clearly was in no way intimidating, as was proved in the Guards Armoured Brigade Patrol Competition, which was won by the 'A' Company patrol, led by Lieutenant P. P. Rawlins, with Lieutenant Chambers' 'C' Company patrol coming third, and the 'B' Company patrol of Second-Lieutenant Wells in sixth place.

Following the usual autumn manoeuvres and 'Fitness for Rôle' inspection, many soldiers managed to get in some skiing in the Harz Mountains or near the Austrian border before the Battalion's next main task was undertaken: Exercise 'Jeremiah'.

This was an exercise sponsored by the Ministry of Defence, but written, run and controlled by the Battalion, to see how well-trained soldiers could cope in a realistic exercise under constant chemical attack, with every activity closely supervised and monitored by Government scientists. The training took place in April, 1972, the first three weeks being spent at Munster, where the troops underwent progressively more demanding tests to judge how the wearing of protective clothing affected the soldiers' performance. The final test consisted of a fourteen-mile run at best speed, wearing respirators and full protective clothing, and carrying weapons and a 51-lb pack. The trials finished with a five-day exercise at Sennelager, with the participants under extensive attack from CS gas. Throughout this unique exercise the Battalion, in addition to coping with the demands of the scientists, received countless visits from VIPs and a large number of establishments. It was generally felt in the Battalion that taking part in the trials themselves, despite their rigours, was preferable to administering the scientists and visitors!

The letters of congratulation and admiration that subsequently flowed into Battalion Headquarters were some recompense for the effort required to mount 'Jeremiah'. No amount of praise, however, could compensate for the fact that, with the exercise ending late on Friday, 28 April, with a very tired Battalion, equipped and organized on the Mechanized Establishment, the troops had the space of a week-end to return to Munster, clean up, rest, and reorganize on to a four-company basis ready to start training on Monday, 1 May for the next tour in Ulster.

The 2nd Battalion's second sojourn in Belfast lasted from July to the first week in December. While the Ulster experience was excellent for the *esprit* of the Battalion and the development of junior leaders, the time spent in the

Province, added to the preparations needed beforehand for the totally different rôle, made serious inroads on the expertise needed in a mechanized battalion, with its 84 APCs, well over 300 radio sets, radar and night vision devices, and the support weapons. Thus, on return to Munster, the Battalion had to start almost from scratch again, training drivers, signallers and other specialists, so that from January, 1973, there was a practically constant figure of 150 men on cadres at any one time. To add to all this, the Battalion had to resume its NCE rôle and prepare for its annual inspection.

On 9 March, 1973, Lieutenant-Colonel Gerrard-Wright handed over to Lieutenant-Colonel M. E. Thorne, whose first task was to take the Battalion, now almost up to strength in specialists, to Soltau for the first opportunity to train with supporting arms since the summer of 1971.

After a most welcome visit from HRH The Duchess of Gloucester on 23 May, the pattern of life in 1972 began to repeat itself in 1973 with preparations for the Battalion's third tour in Northern Ireland, which lasted from July to late November, when once again the cycle of retraining began. In the thirty-one months, from May, 1971 to December, 1973, that the Battalion had spent as mechanized infantry, thirteen months of that time had been devoted to training for, and operating in, Northern Ireland. The nature of operations in Ulster will be described later, but when the diversity of skills required for internal security and mechanized warfare are compared, and the complexities and problems of retraining for both rôles are taken into account, the adaptability, professional expertise and high morale of the modern Royal Anglian soldier must remain a source of pride and wonder.

CHAPTER 8

TWICE A CITIZEN

'Warriors who wore spectacles, and who, off parade, sat peaceably behind their desks or shop counters.'

'The British Volunteers',
Illustrated London News, 27 October, 1860.

So FAR THIS story has concerned itself exclusively with the Regular battalions. However, these only form part of the Regiment, which has three more components: the 5th, 6th and 7th (Volunteer) Battalions, whose absorption into the Regiment must now be traced.

In a Regular, as opposed to a conscript, army all soldiers are volunteers with a small 'v', but a Volunteer with a capital 'V' is the embodiment of a long-standing British tradition: the readiness of many citizens to devote part of their free time, out of normal working hours, to learning the soldiers' trade, and thereby providing a second echelon to the land forces of the Crown.

The threat of French invasion in the eighteenth and early nineteenth centuries produced a host of Fencible and Volunteer units to support the Regular Army, but after Waterloo these were disbanded. In 1859 many influential citizens became alarmed at the growing bellicosity of the Second French Empire under Napoleon III, and steps were taken, despite opposition from Parliament and the Army, to form a force for home defence from public-spirited members of the population. Thus was born the great Volunteer Movement, which resulted in the formation, all over the country, of companies of Rifle Volunteers. Some of these had little military value, their chief attraction in the eyes of their members being the opportunity to wear exotic uniforms and to enjoy the social prestige attached to military rank. Nevertheless, the principle of voluntary service became firmly established in the nation's life, and the idea of part-time soldiering gradually increased in popularity and public acceptance.

The Rifle Volunteer companies were grouped into administrative battalions, which, in 1881, became Volunteer Battalions of the Regular county regiments. In 1908 the Secretary of State for War, Lord Haldane, formed the Territorial Force, which was to provide a second line to the Regular Army, with its units grouped into brigades and divisions like the Regulars. The 'Volunteer' title was dropped and the units became Territorial battalions of their respective regiments. Though originally formed for home defence only, the Territorial Force volunteered for overseas service on the outbreak of war in 1914, and throughout the Great War and the Second World War (having been renamed the Territorial Army in 1920) gave yeoman service to the nation.

After 1945 the Territorial Army remained, in essence, much as it had been before. Though the Territorial battalions formed part of their parent regiments, the TA as a whole was, in law, a separate army under the Crown, recruited and administered, though not commanded, by the County Territorial Associations. The amalgamations of 1958–60 had no effect on it, so that, for example, when the Royal Lincolns and Northamptons became the 2nd East Anglian Regiment, their Territorial battalions continued as 4th/6th Battalion The Royal Lincolnshire Regiment (TA) and 4th/5th Battalion The Northamptonshire Regiment (TA), as did the TA battalions of the other regiments of the East Anglian Brigade. When the latter became the Royal Anglian Regiment in 1964, the seven TA battalions continued under their old nomenclatures.

Then, in July, 1965, the Secretary of State for Defence announced the disbandment of the Territorial Army. In its place would be formed, with effect from 1 April, 1967, the Territorial and Army Volunteer Reserve (TAVR): no longer a separate force, but the reserve arm of the Regular Army. Though many thinking people, including members of the TA and the County Associations, recognized that the TA was in need of reorganization, the announcement, and particularly the way in which it was made – apparently calculated to exclude the views of the Territorials themselves – caused great dismay and distress. The plan produced by the Ministry of Defence, without prior consultation with the Council of Territorial Associations, envisaged a small TAVR with the rôle of supporting the Regular Army in an emergency with units and individuals. After much lobbying in Whitehall, the Government reluctantly agreed that the force should include an element for home defence.

Thus, in 1967, the TAVR was born. Categories I and II, 60,000 strong and known as Volunteers, provided the individuals and units for support of the Regular Army; Category III, known as Territorials, provided a Home Defence Force of 28,000 to support the Civil Authorities at home in time of great emergency. As an indication of the Government's reluctance to form the TAVR III, it should be noted that the force, which consisted of a number of units without any formation headquarters, was to be paid for from the Home Office Vote, and was to be equipped and trained to a lower scale and standard than the old TA.

As far as the Regiment was concerned, the reorganization resulted in the following: as part of TAVR II, a 5th (Volunteer) Battalion was to be formed, recruited from the whole regimental area thus:

Battalion Headquarters, Headquarters Company and Band – Peterborough

No. 1 (Norfolk and Suffolk) Company – Lowestoft

No. 2 (Duchess of Gloucester's Own Lincolnshire and Northamptonshire) Company – Lincoln

No. 3 (16th/44th Foot) Company—Chelmsford
No. 4 (Leicestershire) Company—Leicester.

The Volunteers had to be efficient infantrymen, physically fit, and prepared to go overseas at short notice. Recruits had to do a fortnight's training at the Regimental Depot before they were accepted. They were well-paid and received bounties for training and call-out liabilities. The Battalion had modern weapons and equipment and opportunities for overseas training. With these inducements, the 5th Battalion was soon up to strength, and, in some companies, over-subscribed.

As part of TAVR III, there were to be seven Territorial battalions, each bearing the name of the former TA units, Royal Norfolk, Suffolk and Cambridgeshire, and so on. These battalions had to be content with the old 'battle-dress' uniform, very little transport and equipment, and no weapons other than the obsolete No. 4 Rifle; plus, it might be added, the minimum of official encouragement and direction.

This apparently calculated projection of Category III units as the poor relations of the TAVR failed to deter would-be recruits, and the Government was spurred to sterner action. Exactly one year after the formation of TAVR III, all financial support for the force was withdrawn; no pay, no bounties, no transport; even medical fees for the examination of recruits had to be found by the units themselves. Undaunted and unpaid, the Territorials soldiered on. Money was scraped together from funds and trusts to purchase petrol, and hire transport and meet some out-of-pocket expenses. The 2nd and 4th Battalions, then stationed at Felixstowe and Chatham, and the 5th Battalion all did what they could to help with training. Annual Camps were held, the Bedfordshire and Hertfordshire Regiment even managing an expedition to Guernsey with everything paid for out of regimental funds and individuals' pockets and sustained by enthusiasm. As a Leicestershire Territorial wrote in *Castle*:

'In terms of military achievement, not a particularly memorable year. In terms of frustration and resolution, one that will never be forgotten. Until we are forced by circumstances to abandon all or part of our normal life with the Battalion, business will be as usual.'

'Until we are forced'—this is what it now came to. As a reward for this remarkable manifestation of the voluntary spirit, the Territorial battalions, after a two-year struggle to survive, were ordered to disband on 1 April, 1969. No ceremonies were allowed to mark their passing; no parades to bid farewell to these last bearers of the names of the old county regiments; simply a cadre of three officers and five soldiers was allowed to stay on to keep alive, in each county, the spirit of part-time soldiering. In a Special Order of the Day, the Colonel of the Regiment, Lieutenant-General Sir Richard Goodwin wrote:

'It has been a remarkable reflection upon your loyalty that your spirit has never flagged and that you have cushioned one blow after another, one temporary expedient after another, until the Government at last reached the final conclusions, which we all secretly hoped might be avoided, but which we feared to be inevitable.'

Meanwhile the 5th Battalion, under the command of Lieutenant-Colonel Moore, had got off to a good start. Initially there was the problem of achieving a corporate battalion spirit. The companies were widespread, from Lincoln to Chelmsford, and at the first battalion camp, at Otterburn, all tended to regard themselves as being of their former regiments rather than Royal Anglians. As *Castle* put it:

'It would perhaps be kind to draw a watery veil over the events of the first weekend. Things could have been better but after all we were a new unit. The Companies were all very clannish and self-contained, and groups of company officers were found standing in tight little circles like sheep expecting attack, each group surreptitiously eyeing the other and wondering what the perfidious ——shires would be getting up to in the next two weeks.'

However, within the first four months of the Battalion's existence, Lieutenant-Colonel Moore managed to get all companies together on three occasions, and as they got to know each other, the emphasis began to change from one of mutual suspicion to friendly competition. The Battalion's report to *Castle* summed this up:

'There is obviously going to be far greater inter-Company rivalry than most of us have seen before. This is only to be expected when each Company is trained as an independent sub-unit, and is obviously going to be one of the great strengths of the AVR. Inter-Company feelings ran high whenever there was competition between them, and every company soon developed its own personality and gave the impression of being able to take on the rest single handed.

'Whatever the cynics may say, this is not the old TA with a new name. We witnessed at Otterburn something unique, something with its place in military history. We welded together a new unit from what we believe to be the best fragments of many old ones.'

The companies soon began to achieve a high level of attendance at drill nights and training week-ends. From 1968, companies started to spend their annual camp training abroad. No. 3 Company were the first to go, spending two weeks with a Regular battalion in BAOR, followed in 1969 by No. 2 Company going to Berlin and No. 4 Company to BAOR. The rest of the Battalion assembled for their camp at one of the United Kingdom training areas. Organization of training at camp was, after 1969, greatly assisted by the Territorial Cadres.

In 1969 the Battalion's strength was increased by the formation of No. 5 Company, commanded by Major Keatley, based on Hertford, and later Hemel Hempstead. The following year this company, with No. 1 Company, had the annual camp in Malta, and in 1971 spent two weeks attached to the 3rd Battalion in BAOR. By the end of 1973, Cyprus and Gibraltar had been added to the Battalion's training grounds.

As the Battalion built up, new training centres were opened: at Corby (Headquarters Company), Ipswich (No. 1 Company), Gainsborough (No. 2 Company), Colchester (No. 3 Company), and Loughborough (No. 4 Company). No. 3 Company was affiliated to 19 Airportable Brigade because its mobilization plan allocated it to the 3rd Division; the remainder of the Battalion was designated to complete the Order of Battle in BAOR.

Command of the Volunteer Battalion was vested in a Regular officer, whose Regular staff consisted of a Training Major, Adjutant, Quartermaster, Regimental Sergeant-Major, Orderly Room Quartermaster Sergeant, a driver, and one Warrant Officer Class 2 and one Permanent Staff Instructor per company, with two Staff Sergeants and two Permanent Staff Instructors in Headquarters Company. Lieutenant-Colonel Moore and the Battalion's first RSM, Warrant Officer Class 1 Blood, handed over in 1968 and 1969 respectively; Lieutenant-Colonel Blyth took over and much of the Battalion's subsequent success was due to his efforts to weld together the far-flung companies and to build up a Royal Anglian spirit. He handed over in 1971 to Lieutenant-Colonel Heath, who was in turn succeeded by Lieutenant-Colonel Hopper in February, 1973.

Training at the annual camp was greatly helped by the Cadres of the former TAVR III battalions, who took over responsibility for much of the administration and the setting and running of exercises and competitions. The Cadres' reward came in 1970. Following the General Election, the new Tory Government announced, in November, an increase in the Reserve forces. As a result, the Regiment gained two new Volunteer battalions: the 6th, with Headquarters at Bury St Edmunds, to be recruited from East Anglia; the 7th, with Headquarters at Northampton, to be recruited from the East Midlands. Each battalion was to have a Volunteer Commanding Officer—Lieutenant-Colonel Raywood for the 6th, and Lieutenant-Colonel Wallace for the 7th—a Regular Training Major, Regimental Sergeant-Major and one Permanent Staff Instructor per company, and companies located as follows:

6th Battalion

'A' (Royal Norfolk) Company—Dereham
'B' (Bedfordshire) Company—Bedford
'C' (Essex) Company—Braintree
'D' (Cambridgeshire) Company—Cambridge.

7th Battalion

'A' (Royal Lincolnshire) Company—Scunthorpe
'B' (Royal Leicestershire) Company—Leicester
'C' (Northamptonshire) Company—Northampton
'D' (Northamptonshire) Company—Wellingborough.

In addition, the 7th Battalion was to have a squadron of the Leicestershire and Derbyshire (Prince Albert's Own) Yeomanry, based on Melton Mowbray. The Yeomanry were allowed to retain their own cap badges, but the remainder wore the Royal Anglian cap badge and, as collar badges, each company wore the badge of its former Territorial regiment; this system of badging was also adopted by the 5th Battalion.

The Cadres were disbanded and their officers and men, plus many members of the old Territorial units, could now join the new battalions. The Volunteer element of the Regiment owed much to the Cadres, not only for their assistance to the 5th Battalion, but also for keeping the spirit and tradition of territorial soldiering alive in the counties, particularly in a political climate where there seemed to be forces at work, hell-bent on eradicating the whole concept of voluntary, part-time service, whether it be in the TAVR, Civil Defence, or any other of the many institutions, whose existence depends on the willingness of public-spirited citizens to give their time for little reward.

The two new battalions threw themselves into the task of recruiting, training and doing their utmost to catch up with the standards achieved by the 5th Battalion. The rôle of the new battalions has been stated as 'Home Defence'. For this somewhat vague task they are only armed with rifles and light machine-guns and have no support weapons. A suspicious mind might see in this a sinister similarity to the short-lived TAVR III; the new units do, however, have modern clothing and equipment.

Increasing in strength and military competence each year, by 1973 the battalions were ready, with the 5th Battalion, to receive the accolade of a visit by the Colonel-in-Chief, HM Queen Elizabeth The Queen Mother. The afternoon of 12 July was set aside for the Colonel-in-Chief to inspect all three Volunteer battalions of her Regiment at Bassingbourn Barracks, to see them at work, and to meet as many of her Volunteers and their families as possible. After a most successful day, the Colonel of the Regiment, Lieutenant-General Sir Ian Freeland, wrote to the battalions: 'Queen Elizabeth was particularly impressed with the spirit and keenness of the Volunteers and the good attitude of their wives'.

The disbandment of the old Territorial Army disrupted the often close connections between the county civic authorities and the Territorial battalions. Since the goodwill of the former was important to the whole concept of Volunteer soldiering, the progression from TA to TAVR, and the absorption of the old county TA battalions into the Royal Anglian Regiment, had to be explained and the links between the Regiment and the counties and towns concerned made clear. That this was successfully achieved was manifested by the visits to the Volunteers of Lord Lieutenants, Mayors, Chairmen of TAVR Associations and other dignitaries, and the granting of certain Freedoms: of Peterborough and Hertford in 1969 to the 5th Battalion; of Cambridge in

1972 to the 6th Battalion; and of Wisbech in 1974 to 'D' (Cambridgeshire) Company, 6th Battalion.

So, despite setbacks, the Volunteer spirit lives on and flourishes in central and eastern England. The greens and greys of the Rifle Volunteer uniforms of the 1860s are now merged in the disruptive pattern clothing of the Volunteers of the 1970s; three battalions, part soldier, part civilian, linked to their brothers-in-arms of the Regular battalions by a common cap badge and a common recruiting area.

CHAPTER 9

ULSTER WILL FIGHT . . .

'We are the unwilling
Sent by the unqualified
To do the unacceptable
For the ungrateful.'

Soldier's graffiti in Crossmaglen.

As the Union Jack was hauled down all over what was once the British Empire, and as the Army contracted, many people wondered what exactly, apart from war games in Westphalia, soldiers were going to find to do to justify their existence. A scheme for employing troops as bridge and road builders, known as Military Aid to the Civil Community, gained some prominence. Rhine Army had its well-established rôle; the Strategic Reserve was poised to deal with not easily foreseeable eventualities; garrison battalions stood guard, cleaned their areas, played games and went adventure training. But was there ever again going to be any real soldiering? Barring a nuclear holocaust, which hardly fitted the bill, would soldiers ever have the chance to put into practice their expensive training and, if so, where?

The answer came from a quarter which few military thinkers had even contemplated—their own back doorstep. To the age-old tunes of 'The Wearing of the Green' and 'The Sash my Father Wore', the traditional Irish propensity for violence as a solution to a dispute, erupted. On 12 August, 1969, men of the Prince of Wales's Own Regiment of Yorkshire drove into Londonderry to relieve the hard-pressed Royal Ulster Constabulary and to separate Catholic from Protestant. The Army's longest continuous campaign since the Malayan Emergency of 1948–60 had begun; 'real soldiering' was back with a vengeance—in the United Kingdom itself.

Up to August, 1969, the Northern Ireland Command had seemed a military backwater, its single infantry brigade, the 39th, frequently losing its battalions as overseas reinforcements. By the end of September, with Belfast surpassing Londonderry in devastation, there were ten infantry battalions in Ulster, and the GOC, Lieutenant-General Sir Ian Freeland, late of the Royal Norfolk, had been made fully responsible for security, with police and troops under his command. The arrival of the troops was initially welcomed by the local population, but when General Freeland was asked at a press conference on 18 August how long it would be before the welcome turned sour, he replied:

> 'The honeymoon period for the use of troops in a situation like this can be very short-lived. I think we are probably at the peak of that honeymoon right now. If, for example, something constructive does not come out of the meeting between the two Prime Ministers* in London, then the honeymoon period could finish in a matter of hours. Our soldiers, in between the rival factions in Belfast, might become the target of both sides.'

General Freeland continued in this exacting and often thankless and frustrating enterprise until February, 1971, when he retired with the well-deserved award of the GBE. Seven months later he took over as Colonel of the Regiment from Lieutenant-General Sir Richard Goodwin.

With an officer of the Regiment in overall command, it was appropriate that battalions of the Regiment should play their part to assist him in his most difficult task, a task in which so much depended 'on the confidence the troops could engender in people filled with fear, hatred, suspicion and superstition.'† Battalions in Ulster were of two types: 'resident'

* Mr Harold Wilson and Mr James Chichester-Clark, later Lord Moyola.
† Gregory Blaxland, *The Regiments Depart* (London, 1971), p. 478.

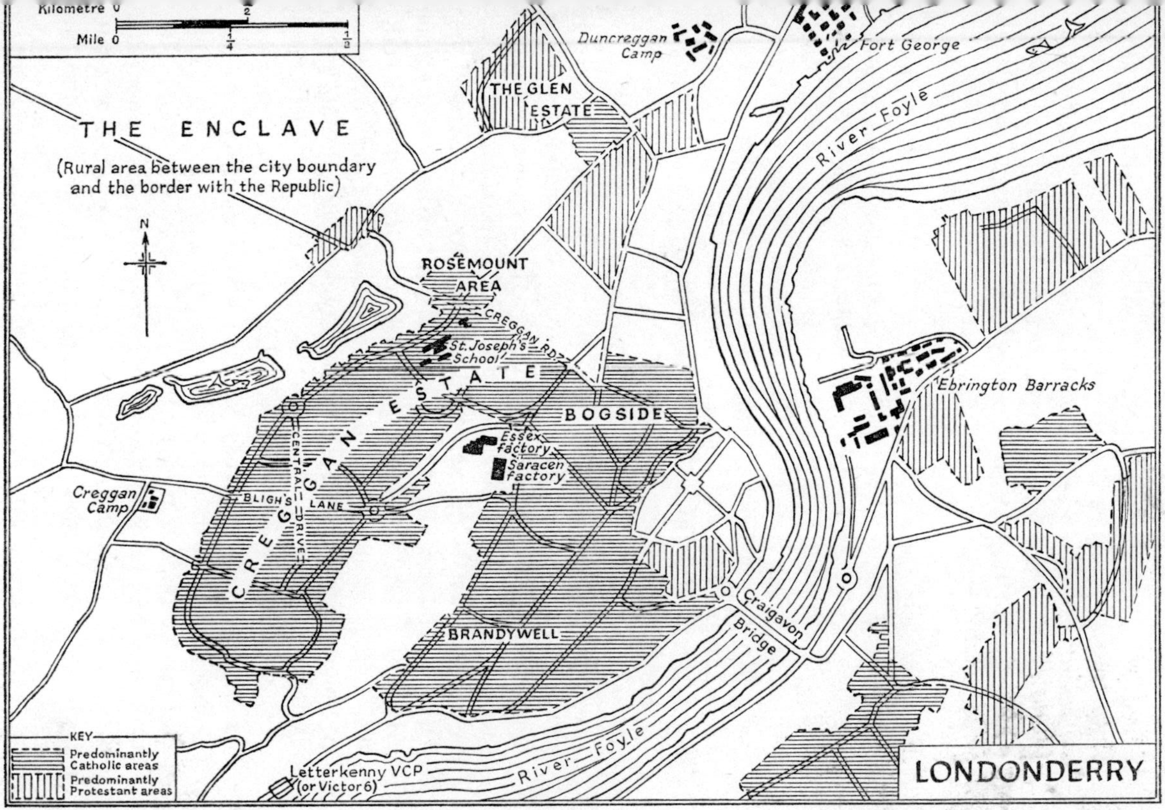
LONDONDERRY
THE ENCLAVE
(Rural area between the city boundary and the border with the Republic)
Kilometre 0 2
Mile 0 1/4 1/2
N
Duncreggan Camp
Fort George
River Foyle
THE GLEN ESTATE
ROSEMOUNT AREA
CREGGAN RD
St. Joseph's School
CREGGAN ESTATE
BOGSIDE
Essex Factory
Saracen Factory
Ebrington Barracks
Creggan Camp
BLIGH'S LANE
CENTRAL DRIVE
BRANDYWELL
Craigavon Bridge
Letterkenny VCP (or Victor 6)
River Foyle
KEY
Predominantly Catholic areas
Predominantly Protestant areas

battalions, which came with their families on a two-year tour, and those on four-month emergency tours from their permanent stations in England or Germany. The 1st Battalion was of the former type, and the 2nd and 3rd of the latter, completing five such tours between them from 1970–1973.

1st Battalion, 1970–72

Before leaving Catterick and the mechanized rôle, the Battalion reorganized for its new and very different undertaking. Battalion Headquarters was split into Tactical and Main HQ. The three Rifle Companies and Support Company were formed into four Internal Security (IS) Companies, each of Company HQ and three IS Platoons of a minimum of twenty-four all ranks. The Battalion was completed by the Reconnaissance Platoon of five sections; the Defence Platoon (Drums), also of twenty-four all ranks, so that it could take the place of any IS Platoon; Numbers 13 and 14 Platoons, made up of soldiers from Command and Administrative Companies, not employed on essential duties; and the Band, all of whom were trained as medical orderlies, two bandsmen being deployed to each IS Platoon and the Defence Platoon, the remainder assisting at the Regimental Aid Post. Thus arrayed, the 1st Battalion left Catterick and, under the command of Lieutenant-Colonel Jackson, was complete at Ebrington Barracks, Londonderry, by 30 July, 1970.

On arrival it was found that there was insufficient accommodation in the barracks for the whole Battalion, so 'C' Company had to move across the River Foyle into Duncreggan Camp, last used by the Regular Army in 1935! The Company barely had time to register its feelings about its accommodation, for, at 22.00 hours on 8 August, it was put on immediate standby. Ten minutes later it was deployed in the Bogside, as its correspondent to *Castle* described:

'With two Saracens in support from the 17/21st Lancers and

9 Platoon up, we had to crash through a blazing barricade made up of cars, vans and debris, etc., with the rioters throwing missiles from the other side. Permission from the CO was given to use CS gas. We fired 35 rounds, broke through the blazing inferno and drove the rioters back into the Bogside.'

The early part of 1970 in Londonderry had been marked by mass rioting and wanton destruction, but early in the Battalion's tour this had tailed off into sporadic and predictable rioting with hooligans throwing stones, bottles and the occasional petrol bomb. However, the situation was lively enough for the Battalion to be the first to employ rubber bullets—or baton rounds as they were officially termed—against rioters in August, and shortly afterwards part of the Battalion found itself containing rioting Catholics in the Bogside, while 'B' Company fired CS gas at a horde of Protestants on Craigavon Bridge.

Despite such outbreaks, tension continued to ease in Londonderry to the extent that the Bogside and Creggan were placed 'in bounds' to off-duty soldiers. Schemes to win back the confidence of the people were set in train and met with considerable success—to the chagrin of the IRA.

Although things were quiet in Londonderry at the start of 1971, the Belfast pot was boiling over. In February the Battalion was despatched to reinforce the hard-pressed troops there, deploying to search the Clonard, and taking over part of the operational area of the 2nd Battalion, which had arrived in Belfast in October, 1970. The 1st Battalion was struck by the difference between Londonderry and what it found in Belfast, recording in *Castle*: 'the simmering hatred on both sides; the vicious abuse, particularly from the women; and the abundance of shots and nail bombs. Although we did not realize it at the time, this was a grim foretaste of what was to come in Londonderry.'

During the Battalion's absence, rioting started again in Londonderry, and on returning, 'A' and Support Companies went straight into action, sweeping through the Bogside and making twenty-two arrests. By this time the IRA were reacting to counter the success of the 'hearts and minds' campaign, and in May, Private Larter was shot in the hand while escaping from three armed men in the Bogside, who had set upon him as he was saying goodnight to his fiancée; his was the first gunshot wound to be received in Londonderry, but many more were to follow.

After this incident the Bogside and Creggan were placed out of bounds and in July, after block leave, the Battalion returned to a sudden, deliberate attempt by the IRA to escalate the violence, by opening fire on the troops. After a year in which the Battalion had only been attacked with hand-thrown missiles, in the space of a few days more than 140 rounds had been fired at members of the Battalion. On 8 July the troops opened fire for the first time and an armed man and a nail bomber were both accounted for.

The relatively peaceful months of the Battalion's first year were over and were not to return. Stones and bottles gave way to bombs and bullets; internment was introduced; the barricades went up in the Bogside, rioting became more vicious, and the Battalion began to suffer its first casualties, as the IRA tried every device to get soldiers on to the streets as targets. In August Lance-Corporal Turner of Support Company was wounded in the head by a ricochet while removing a barricade. The Battalion levelled the score when three riflemen engaged a gunman armed with a pistol at 30 metres range; the date, unluckily for the gunman, was Friday, 13 August! In September an isolated post at the Mex Garage in the Brandywell part of the city came under constant attack from stones, bullets, petrol and gelignite bombs. On the 27th, Private Wilkins was on duty in the observation post when over fifty rounds were fired at his position. He died from his wounds, the Battalion's first fatal casualty from enemy action since Aden. On 9 November, during an anti-mining patrol, Lance-Corporal Curtis was hit in the chest while crossing a gap in a wall. One of the patrol gave covering fire, while the third man went to the aid of the stricken NCO, but Curtis died on the way to hospital, twenty-four hours before the Battalion was due to go on leave.

The Battalion now knew that it was to leave Ulster in March, 1972, and returned from leave to its last tour of duty in the city of Londonderry, to face more sniper attacks and a sombre Christmas, as *Castle* described:

> 'Pre-Christmas in the city was eerie. While shoppers bustled busily about buying gaily wrapped presents under colourful festive lights, patrols eased themselves through the throng suspicious of every parcel and any bulky coat. As the afternoon drew on, hooligans would gather on the fringe of the Bogside and, bored by the lack of activity, would make brief sporadic sorties to throw stones into the preoccupied crowds. Troops would then be hurriedly deployed so that Christmas for the locals could be as normal as possible. Then, until the hooligans got tired or the shops closed their doors to customers, the soldiers would stand shivering in darkened streets peering round draughty corners looking for the gunmen or the sneak nail bomber. The sudden crack of a bullet or explosion of a nail bomb would tell them that somewhere nearby "somebody cared". Fifty yards away the shoppers, hearing the noise, would give a frightened glance then continue slightly more hurriedly on their Christmas spending spree. Those of us a bit further back would say a brief prayer and give a sigh of relief when the report of "No cas" came back on the radio.'

On 21 December the Battalion withdrew from the city to revert to the rôle of policing the county, and its own Christmas. It took with it a ship's cannon, dating from the siege of 1688, presented by the Committee of the Apprentice Boys in recognition of the Battalion's work in the city.

The county rôle, though less lively, still required three companies permanently deployed, with the fourth at notice as brigade reserve. Throughout the tour, the families, many of whom for some time had had to be accommodated up to forty-five miles away from the Battalion, had given every support to their hard-pressed menfolk. When an empty

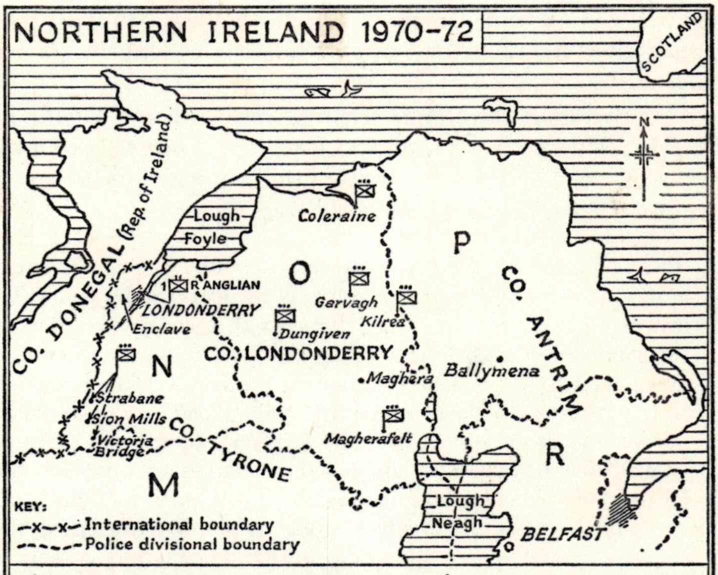

1st Battalion deployment in the 'County Role', showing main locations manned by platoon size subunits

married quarter was blown up, the security of the families became yet another task for which men had to be found. Despite the difficulties this entailed, the Battalion could at least count the days to its departure, look forward to months in the sun of Cyprus, and reflect, with some pride, on the parting words of the Commander Land Forces, Major-General Ford, who told them that, despite the limitations imposed on offensive operations, their spirit, high standards and professionalism had been maintained, which was an achievement worthy of the highest praise.

Some statistics from the Battalion's twenty months in Northern Ireland provide some interesting sidelights on the operations. Out of 996 shots fired at the Battalion, only ten were fired in the period August, 1970–June, 1971; all the 226 shots fired by the Battalion, which resulted in twenty-nine gunmen being seen to fall or be hit, were discharged between July, 1971 and March, 1972; 2,698 rubber bullets were fired at rioters, of which only sixty were fired in the period up to June, 1971; the operations cost the Battalion twelve casualties, of which two were fatal, and all save three occurred in the period from July, 1971 to the end of the tour. These figures not only show how the violence escalated dramatically from July, 1971 onwards; they also indicate, first, how firmly the Battalion abided by the principle of minimum force, and secondly, the forbearance and endurance shown by the soldiers over a long period of stress. For his command of the Battalion, Lieutenant-Colonel Jackson was granted an operational award of the OBE. In the words of the citation, his 'personal example and calm and fearless leadership' had been 'of a very high order, often under exceptionally difficult and trying circumstances, throughout sixteen arduous months.'

2nd Battalion, 1970–71

As mentioned earlier, the 2nd Battalion arrived in Belfast from Colchester on 14 October, 1970. Although it had, for a time, responsibilities for part of the city east of the River Lagan, its main concern was West Belfast, from the mainly Protestant Springmartin area to the north, across to the Catholic Clonard to the east, then south through Ballymurphy and Andersonstown, both strongly Catholic.

There had been serious riots in Ballymurphy in April of that year, following which recruitment to the IRA grew apace. The estate had a high crime and unemployment rate, added to an inborn hatred of the Army and the RUC. This volatile mixture erupted in further rioting shortly after the Battalion's arrival, and it fell to 'B' Company, with assistance from 'C' Company, to restore order.

The area remained simmering, however, and, in an effort to forestall more trouble, Lieutenant-Colonel Gerrard-Wright started negotiations with the local Provisionals. *The Sunday Times*, in its best 'all things to all men' vein, described this venture thus:

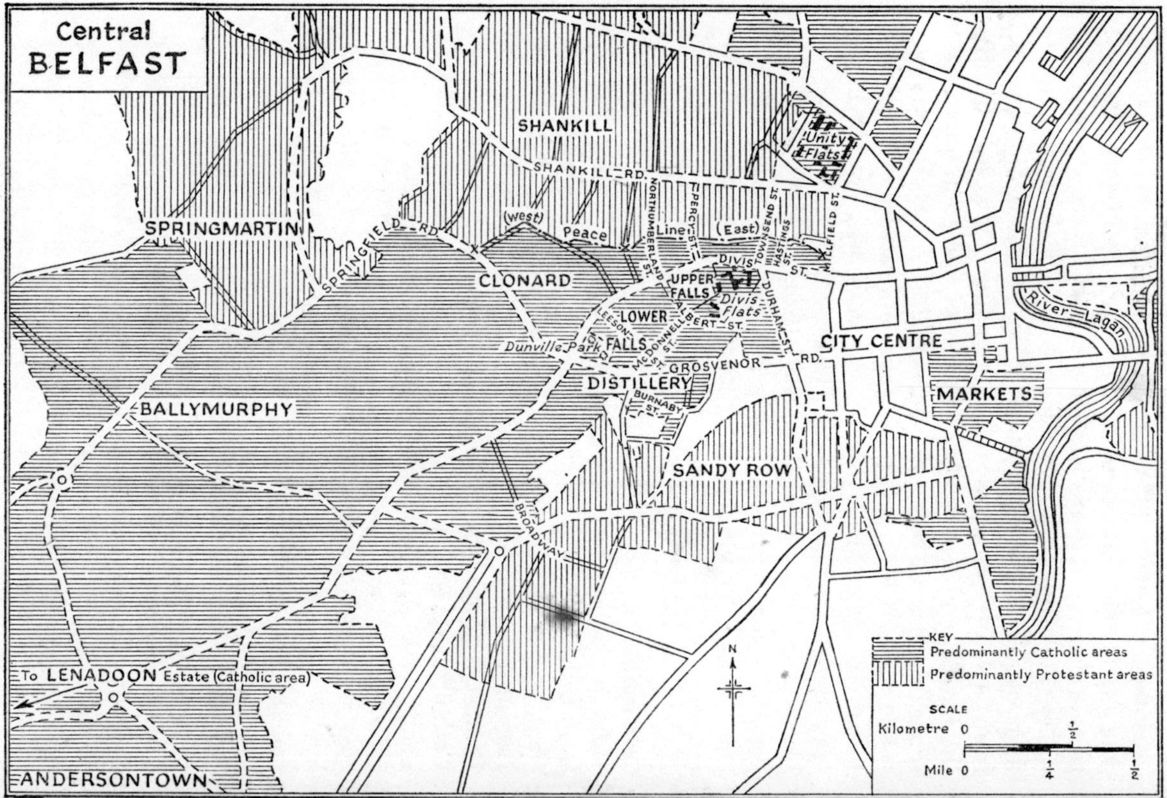
Central
BELFAST
SHANKILL
SHANKILL RD.
SPRINGMARTIN
SPRINGFIELD RD.
(west) Peace Line (East)
NORTHUMBERLAND ST.
PERCY ST.
TOWNSEND ST.
HASTINGS ST.
MILLFIELD ST.
Unity Flats
CLONARD
DIVIS ST.
UPPER FALLS
Divis Flats
LOWER FALLS
LEESON ST.
McDONNELL ST.
ALBERT ST.
DURHAM ST.
Dunville Park
GROSVENOR RD.
DISTILLERY
BURNABY ST.
CITY CENTRE
MARKETS
River Lagan
BALLYMURPHY
SANDY ROW
BROADWAY
To LENADOON Estate (Catholic area)
ANDERSONTOWN
N
KEY
Predominantly Catholic areas
Predominantly Protestant areas
SCALE
Kilometre 0 1/2
Mile 0 1/4 1/2

'Although the Anglians are a notably unflappable regiment, the help of the community leaders was clearly needed to keep the peace.*

Despite the Commanding Officer's great personal efforts to attain a *modus vivendi*, Ballymurphy once more exploded on 11 January. Even had they wished to do so, the Provisionals were incapable of controlling the thousands of teenage hooligans which the estate contained.

A reporter, who was at 'B' Company Headquarters, described the onset:

'At midnight we heard distant, rhythmic signalling by dustbin lids, an eerie sound taken up around a drab, postwar housing estate that is the greater part of Ballymurphy. Five minutes later what sounded like an army on the march could be heard approaching. There was clanging of more bin lids, and nationalist songs that changed into ear-shattering abuse. A crowd of about 50 men started tearing up the pavement for ammunition. Long sections had already disappeared in previous confrontations and finding missiles was easy.'†

Full-scale rioting ensued; cars were set alight, petrol bombs and sulphuric acid followed the paving stones, all to the accompaniment of obscene insults from screeching women.

To disperse the mob, 'B' Company fired CS gas and rubber bullets and deployed what was to prove the most effective deterrent—their 'Ham Squads'. Led by such stalwart NCOs as Sergeant Travis, who played in the second row for Felixstowe Rugby Club, Corporal Hampton, who had formed the first of these squads, and Corporal Sanderson, these groups of big, strong men charged again and again into the thick of the most violent sections of the mob to arrest the ringleaders, a task demanding toughness and courage.

Throughout these operations Captain Hart, selected at short notice to command his company, displayed outstanding

* *Sunday Times* Insight Team, 'Ulster' (London, 1972), p. 236.
† *Ipswich Evening Star*, 18 January, 1971.

bravery and coolness. He was awarded the MBE for gallantry and the citation recorded:

> 'A particularly ugly situation developed during a search operation in the heart of Ballymurphy on 15 January. He personally supervised the extraction of nine search teams—three of which he brought out himself. His brave, prompt and bold action ensured a successful conclusion to an operation which might otherwise have ended far less happily.'

The Reconnaissance Platoon, commanded by Sergeant Nugent, operated squads designed to outflank rioting crowds. On 14 January two barricades had been built and set alight by rioters, whom it was suspected were using the area as an ambush position. Sergeant Nugent mounted a quick but carefully planned operation designed to take on the crowd from four different directions, with himself acting as decoy. As expected, it proved to be an ambush and he was viciously attacked, being quite badly burned by a direct hit from a petrol bomb and only narrowly escaped a savage swing from an axe. The crowd was dispersed and all his men were safely extricated under his supervision. For this gallant action and many similar operations he was awarded the BEM.

Although the Battalion could produce a robust response when required, its soldiers were equally capable of many acts of humanity and kindness to the local population, even when having to contend with a riot at the same time. Company Sergeant-Major Hewitt went to the aid of a woman having an epileptic fit in Ballymurphy, and some other 'B' Company men saved two women, one elderly and dying, from a threatening group. Much work was done for youth clubs, parties were organized for old-age pensioners and children, and carol services were held in hospitals at Christmas. As Lieutenant-Colonel Gerrard-Wright told a reporter:

> 'Royal Anglian soldiers are particularly well-equipped (in the Ulster situation) because of their innate good humour, tolerance and steadiness.'*

* *Ipswich Evening Star*, 18 January, 1971.

After the Ballymurphy riots, and with only two weeks of the tour to go, 'C' Company in the Clonard was the next into the fray. *Castle* recorded that the Company's area along the Peace Line was 'a depressing slum area, full of dirty children, stray dogs and littered with barbed wire barricades dividing Catholic and Protestant'. There had been rioting there early in the New Year with the Shankill Road Protestants flexing their muscles. A series of incidents followed but trouble really broke on 3 February. On that day a cordon and search was mounted in the Clonard to investigate three suspected arms caches. This sparked off two days of rioting in which, for the first time, firearms were used against the Battalion. 'C' Company held their ground until help came in the shape of 'B' Company of the 1st Battalion, a company of the Royal Regiment of Fusiliers, and part of Tiger Company under Second-Lieutenant Field. With the end of these riots came the advance party of 3rd Battalion The Parachute Regiment and the end of the 2nd Battalion's first tour.

Comparing soldiering in England with that in Belfast, an officer of the Battalion wrote in *Castle*:

> 'Our whole attitude is changed. We feel ourselves soldiers first, civilians as a hobby. Apart from this fundamental change of attitude, there are other aspects of our life here which have done us nothing but good. For instance, the Battalion has a much better corporate spirit, feeling that we all belong to the same mob, that we can take this job by the ears, pick it up, shake it and spit it out and move on to the next one—in short we have become more professional.'

3rd Battalion, 1972

The 3rd Battalion left Paderborn for its first tour in Belfast on 11 April, 1972, and the next day took over the Catholic Lower Falls from 1st Battalion The Gloucestershire Regiment. The Battalion area was bounded on the north by the Protestant Shankill Road and on the south by Grosvenor Road, the western boundary being the Upper Falls and Springfield Road, and the eastern, Millfield and Durham

Streets. The line of the Falls Road—Divis Street divided the area in half. 'B' and Support Companies were in the north, with the former responsible for the Clonard and the Peace Line West, and the latter for the Peace Line East: 'A' Company looked after the Lower Falls and the Divis Street flats, 'C' Company the Distillery and the rest of the Lower Falls; Battalion Headquarters was in Hastings Street. Part of the area consisted of light industries or business premises, many of which were disused; the larger part was simply street after street of mean, squalid houses, thousands of them derelict, looking, with their gaping windows and doorways, like rows of skulls. Looming menacingly like a fortress over the 'A' and Support Company areas was the Divis Street flats complex, with its single tower block brooding over all.

When the Battalion arrived, the area was quiet, following the Government's political initiative made shortly before. The entry of the Pompadours seemed almost to spark off trouble, for within their first forty-eight hours there were seven shootings, an armed robbery, three bomb hoaxes and two crowd situations to be controlled. On 14 April, Joe McCann, a leading IRA Official, was shot by the Security Forces, and on the 15th the nine-day battle of the Divis Street flats began, with 'A' and part of Support Company in the thick of it.

The Battalion had to contend with mass rioting by mobs of hooligans from the flats, hi-jacking and burning of vehicles, blast and nail bombing, the erection of barricades, and considerable shooting from the IRA. On 16 April rioting went on for nearly ten hours; around 6 pm, when the mob's hysteria was at its peak, 'A' Company came under fire from the flats and Second-Lieutenant Hull was shot, later dying of his wounds. On the next day, Second-Lieutenant Chisnell was seriously wounded while investigating a bomb scare near the Falls Road. Support Company positions were now also under fire, and 'C' Company was dispersing crowds in the Distillery.

The rioting and shooting went on. On the 19th 177 shots

were fired at the Battalion. During the afternoon Lance-Corporal Farmery and Private Lawrence of Support Company recovered a hi-jacked lorry under automatic fire, which shattered the windscreen as they were getting it away. The various Battalion bases formed a ring around the flats and, as the Battalion diary noted:

> 'The battle took on a mediaeval aspect; the rioters, gunmen and bombers sallied forth from their "fortress" in the Divis Street flats to do battle with the Security Forces who, in turn, sallied forth from their fortresses in Hastings Street, Albert Street Mill and the Support Company bases in Townsend, Percy and Northumberland Streets. The Hastings Street base was particularly suitable for this as it adjoined Divis Street, and the complex of buildings and derelicts which lay between Hastings and Townsend Streets was converted into a warren of interconnecting passages and positions from which our own snipers could operate and from which snatch parties could debouch into the fray.'

Although mobile patrols and quick reaction forces contained the disorder at a certain level, the enemy fortress provided a safe sanctuary for the howling mobs, who scurried for safety when hard-pressed, but reappeared in force when the streets were clear of troops. Lieutenant-Colonel Hall-Tipping realized early on that the solution lay in the occupation of the flats, but permission to do so was refused until 21 April. On that day 'A' Company moved into the Catholic stronghold and, after five arrests were made, the rioting quietened down. 'A' Company evacuated the flats but moved back again the next day when rioting recurred. Thereafter it began to diminish as the parents of the young hooligans started to exercise some much-needed restraint on their offspring.

The gunmen were still busy, however, two soldiers being wounded by a burst of automatic fire from a moving car into the back of the armoured vehicle in which they were travelling. On the 25th three soldiers were wounded in separate incidents; one, a section commander, fell in Burnaby Street and was rescued owing to the quick and courageous action of Private Holmes, who had but recently joined from the Depot. By 26 April 559 shots had been fired at the Battalion in the space of a fortnight.

With the situation in Divis Street calming down, the Battalion's area increased to include the City Centre, where car bombing was now on the increase, the Protestant Sandy Row, the Catholic Markets and Broadway, and a large mixed neighbourhood in South Belfast. Support Company, 1st Battalion The King's Regiment, based on the Broadway, was placed permanently under Lieutenant-Colonel Hall-Tipping's command on 13 May, and in the weeks that followed the Battalion had eight or nine companies under command on any one day.

The pattern of action in May included minor trouble from the flats, an increasing number of confrontations with mobs in the Broadway and Distillery, control of movement in the City Centre to counter the car bombers, the supervision of sectarian marches and the surveillance of Catholic funerals. The Battalion's biggest success lay in the City Centre, where the IRA bombs were slowly bringing the economic and social life of the city to a halt. By implementing a large 'Pedestrians only' area in the City Centre, the efforts of the car bombers were diminished and confidence began to return to the people in the streets and offices.

Gun-battles continued throughout the month. The Support Company bases were frequently under sniper fire and 'C' Company also suffered several shooting incidents, losing Private Ballard, who was killed on 11 May when on patrol in the Lower Falls. May was also marked by the setting up of barricades by the Protestants, who were becoming exasperated by the apparent condoning of the 'no-go' areas in Londonderry and the release of large numbers from internment following the political initiative. This culminated on 21 May by widespread barricading throughout the city by Protestant gangs, in the course of which an attack was made on Support Company, who nevertheless succeeded in dismantling the barricades.

On 8 June Corporal Jephcote of 'B' Company saved his patrol from ambush by an intelligent appraisal of the atmosphere in the area at the time. He managed to locate the gunman, charged him and prevented him from taking offensive action against the patrol. Jephcote repeated this tactic the following day. Later a search revealed a cache of arms and ammunition. On the very next day Corporal Jephcote again distinguished himself by arresting a man who was found to be carrying ammunition. For his quick thinking and courage, this NCO was awarded the Military Medal.

The Official IRA had declared a cease-fire on 28 May and on 26 June the Provisionals followed suit. Any advantage accruing from this, however, was counterbalanced by the Protestant UDA, who started to set up their own 'no-go' areas. As a result, tension between the Army and the UDA increased, leading to a massive confrontation on 3 July, in which a clash was only averted by the personal intervention of the Commander, Land Forces.

Meanwhile the Provisionals had used their cease-fire to regroup and rearm and, on 9 July, a stage-managed incident on the Lenadoon estate between the inhabitants and troops gave the IRA the excuse to resume hostilities with extensive shooting throughout Belfast.

On 12 July Lance-Corporal Rooney was killed by a sniper and on the next day Corporal Mogg was shot dead near Dunville Park. In the evening a fierce gun battle raged along the Peace Line for several hours, with Protestant gunmen joining in the fray, in which the Battalion fired 1,344 rounds.

The resumed Provisional activity, which led up to the twenty-six bomb outrages in and around the City Centre on 21 July, afterwards known as 'Bloody Friday', demanded an increasingly aggressive military response. Patrols and ambushes were mounted nightly in areas where the gunmen lurked, and between 21–24 July the Battalion mounted three search operations, but the results were disappointing; the terrorists appeared to have had prior warning. Much the same applied on 31 July, when the Security Forces moved in strength to dismantle all the 'no-go' areas in Londonderry and Belfast in the operation known as 'Motorman'. The barricades came down but most of the birds had flown.

'Motorman' brought to an end the 3rd Battalion's Belfast tour and, after four months packed with incident and casualties of four killed and thirty-six wounded, the Pompadours were not sorry to hand over to their relief—the 2nd Battalion.

2nd Battalion, 1972

The hand-over, which took place on 2 August, was the first time that one battalion of the Regiment had relieved another. For the 2nd Battalion, this new area of the Lower Falls was much smaller than that of the earlier tour, but against this it was totally Republican and hostile. Since the 3rd Battalion's arrival, it had grown progressively more run-down, even the modern Divis flats now having a slum aspect with all the lifts and most of the stairway lights smashed. In this unsavoury atmosphere the IRA enjoyed total support, either through belief or intimidation.

A fortnight after the Battalion's arrival, it suffered its first fatal casualty when Corporal Boddy of the Drums was killed by a sniper's bullet on 17 August while leading his section in the Lower Falls. In September two more soldiers lost their lives. On 23 September Corporal Barry of 'A' Company was shot in the head while leading his section in McDonnell Street and died of his wound two days afterwards. A week later Private Burt was killed by a sniper while on patrol in Albert Street; he had only joined the Battalion in June, having enlisted as a junior leader in 1969.

The IRA was not having it all their own way, however. In September Corporal Sanderson's section ambushed a gunman in Albert Street, killing him and securing his rifle. Corporal Wright hit another terrorist at 600 metres range from an observation post in the Divis tower. During the whole tour the Battalion accounted for five terrorists killed and put away nearly fifty more.

The biggest coup of all occurred in mid-September. During one afternoon Corporal Nicholls of 'A' Company was patrolling with his section mounted in a Saracen APC, when he spotted a stolen car in a queue of traffic. Seconds later the two vehicles met in Leeson Street, where the wanted car reversed wildly, hitting a wall. As one of its two passengers raised a gun, Private Sentance, the Saracen gunner, ordered the driver to ram the car. The two passengers fled, pursued by Corporal Nicholls' section which exchanged shots, hitting one of the gunmen who disappeared. The section was now joined by the rest of 3 Platoon, 2 Platoon and 7 Platoon of 'C' Company. Second-Lieutenant Groom and Sergeant Arnold followed a trail of blood, which led from the yard of one house, over a twelve-foot wall into another house, where they found the wounded terrorist being given first aid by three others, all of whom were seized. During this chase a cut-off group of Sergeant Whitfield and Lance-Corporal Codling, while climbing a wall, suddenly came upon one of the most wanted men in Belfast pointing a pistol at another foot patrol. The terrorist, hearing their movement behind him, ran off but was confronted by Lance-Corporal Ley and Private Amberton. Ley brought him down with a rugger tackle and the man was taken.

Sergeant Whitfield was one of the most prominent NCOs of the Battalion; he continued to be an inspiration to the younger NCOs and soldiers and was responsible for several highly successful actions against terrorists. His qualities were recognized by the award of the DCM, the citation for which stated: 'He has displayed outstanding courage and bravery which was backed by exceptional leadership and professional ability.'

The MBE for gallantry went to Major Barnes for his 'bold and fearless' leadership of 'A' Company in the extremely hostile Lower Falls, where his imaginative and competent handling of many dangerous situations so inspired his men with confidence that terrorist activity in the area was seriously impaired.

Other successes achieved by the Battalion during the tour were the finding of thirty-seven weapons and large quantities of ammunition and explosives. The Intelligence Officer, Captain Howe, pursued every conceivable lead in hunting down the terrorists and it was very largely due to his personal efforts that the Battalion achieved such skill in dominating the area. He too was awarded the MBE for gallantry. On the debit side, the Battalion endured 262 shooting incidents over a total period of 130 days, which resulted in twenty-one soldiers wounded and one more fatality, Private Mason of 'C' Company, who was killed by a sniper on 24 October while on patrol in the Lower Falls. He was a young soldier who had only been with the Battalion for six weeks.

In a foreword to the Battalion magazine, *The Poacher*, written at the end of the tour in December, 1972, Lieutentant-Colonel Gerrard-Wright said:

> 'Both the tangible and intangible results merit the very highest praise and I want here to direct that praise towards those who most deserve it—the Platoon Commanders, Platoon Sergeants, Section Commanders and Private Soldiers. *They* are the people who have achieved the results. It is they who have worked almost impossible hours in a very dangerous environment, who have been constantly abused, insulted, assaulted and shot at. *They* are the people who have borne the brunt and have once again proved how incomparable the British soldier, and in particular the Royal Anglian soldier, is. And throughout the tour the whole Battalion has remained alert, high spirited and incorrigibly cheerful which has been an immense source of encouragement.'

Two further awards of the MBE were made to officers of the Regiment during the period September, 1970–December, 1972. Both were for intelligence duties. Major Bradshaw had overall responsibility for intelligence liaison in the greater

Belfast area which involved the provision of intelligence to a Brigade headquarters and up to ten major military units spread over five police divisions. He made a significant contribution in establishing excellent relations between the Army and the RUC and for most of the time worked in continuing personal danger. Captain Thompson, who won an MC in Borneo in 1964, was Brigade Intelligence Officer at 39 Infantry Brigade. His citation recorded that:

> 'He has at times driven himself to the limits of his strength and totally disregarded his own personal comforts and conveniences. He has played one of the major rôles in the fight against the Provisional IRA.'

3rd Battalion, 1973

Nearly four months after the departure of the 2nd Battalion, the 3rd moved again to Ulster, this time to Londonderry, taking over from 1st Battalion Grenadier Guards on 27 March, 1973. Its main area of responsibility was the Creggan Estate, to which was added the Rosemount area and the Glen Estate to the north, and the rural area around the south of Londonderry known as the Enclave, between the River Foyle and the Republic's border to the west.

The Creggan was totally Roman Catholic and Republican, with 15,000 people living in a postwar housing estate with one public house and one set of shops. It had a very high ratio of young people, who had little to do other than to harass the military. The inhabitants were very aware politically, more so than those of Belfast, and quick to complain about the Army—to the Secretary of State in person, if need be. The estate had not been effectively policed for four years and intimidation by the IRA successfully precluded any likelihood of information being passed to the Security Forces.

The Battalion deployed with 'A' Company, responsible for North Creggan, in Bligh's Lane; 'B' Company, responsible for South Creggan, and 'C' Company, responsible for the Enclave, were in Creggan Camp to the west of the estate, and

Support Company, responsible for Rosemount and the Glen, in Brooke Park.

The Battalion's arrival was greeted with four days of rioting in and around the estate. This continued in varying degrees for the first two weeks, after which a slackening of hooliganism was replaced by shootings at patrols and bases, during which, on 27 April, Private Goodfellow was killed by a gunman while manning the Tesco Vehicle Control Point on the Creggan Road. Private Goodfellow had been a Queen's Scout and had achieved a Gold in the Duke of Edinburgh's Award Scheme. His mother made a remarkable gesture by flying to Londonderry to present to the Scouts of that city a cup for their annual orienteering competition, donated by the Scouts of Rushden in memory of her son.

In the middle of May there was further serious rioting in the Creggan, and on the 23rd a Saracen of the Battalion was engaged by terrorists with a rocket launcher in Central Drive—unsuccessfully as it turned out. The local elections at the end of May passed off relatively smoothly but the ensuing period, up to the Assembly elections on 28 June, was marked by increasing terrorist activity, with rocket, rifle and mortar attacks on Bligh's Lane and Creggan Camp.

The elections passed off with only one attack on a Battalion patrol, but thereafter the violence continued as before. On 4 July Bligh's Lane and Creggan Camp were attacked almost simultaneously with rockets. A running gun battle followed right across the estate in which 200 rounds were fired by the Battalion. One gunman was hit and taken to hospital, and four days later a nail bomber was shot and subsequently arrested. During this battle Lance-Corporal Simpson's section came under fire at a range of 250 metres while advancing down a wide, open thoroughfare. With complete disregard for his own safety Simpson worked his way down the street until he could engage the gunman with success. A month earlier he had led his section in a successful charge against gunmen, thus saving another section, which was unavoidably caught in the open, from suffering casualties.

He was awarded the MM for what the citation described as 'courage and leadership far above that expected of one of his rank'.

At the end of its tour, the 3rd Battalion had carried out a rather more aggressive policy than normal, in order to conform with directives issued by superior headquarters and requests from the police. This policy ensured that the Security Forces maintained the initiative and nipped any serious rioting in the bud. A few block searches were also authorized; these produced some good intelligence but regrettably no arrests. Identification of terrorists at section level became a feature of pre-tour training and much useful information was passed between battalions.

In the middle of July the 2nd Battalion's advance party arrived to take over. The main body arrived at the same time as a reversion to the 'low profile' policy was ordered and the level of violence dropped. On 25 July the 3rd Battalion started to move back to Germany.

2nd Battalion, 1973

The 2nd Battalion, now under command of Lieutenant-Colonel M. E. Thorne, took over on 27 July. The Companies were deployed throughout the Creggan from north to south in the order, Support, 'C', 'B', and 'A', with the headquarters of the first two in the Bligh's Lane base, which included the Essex factory, and those of 'A' and 'B', with Battalion Headquarters, in Creggan Camp; the Reconnaissance Platoon looked after the southern half of the Enclave.

The Battalion was plunged into furious activity in its first two weeks as the Provisionals sought to test the mettle of the newcomers. One rapid success was achieved. Before arrival, much time had been spent training section identifiers; as a result, within four hours of taking over, a man was recognized by Private Saunt of the Anti-Tank Platoon, was arrested and later charged with murder. During the rest of the tour a further five hard-core Provisionals were picked up through the alertness of the identifiers. To help dominate the estate, the Battalion established nineteen covert observation posts, often hidden in the roofs of houses and flats. By the end of the tour twelve of these were still uncompromised.

Lieutenant-Colonel Thorne discontinued the block searches, which in the Creggan seemed unproductive, and all 2nd Battalion searches were related to information received. He also made clear to the locals that arrests would continue to be made following rioting and, by the end of the tour sixty-three such arrests had been effected. However, in order to achieve something that might encourage the flow of information, a system was introduced, following arrests, of warning rather than always charging the offenders; this system only failed once.

In addition to these new ploys, the usual methods of patrols, overt observation posts and vehicle control posts were used to supervise the estate.

The most trouble in the first three months was in 'B' Company's area, where initially there was rioting and some shooting every day. Under the vigorous command of Major Drummond, the Company's tough and disciplined reaction to this violence led to a gradual reduction in its frequency. On 31 July, a day of heavy rioting, Colour-Sergeant Underwood's 5 Platoon made seven arrests after a surprise charge on the rear of the mob; later the Platoon rescued two buses which had been hijacked. For his action this day, and for consistently high standards of leadership throughout the tour, notably in another action on 11 September and for his successes in capturing an explosives expert, bomb-making kit and documents, Colour-Sergeant Underwood was subsequently awarded the MM. The achievements of 'B' Company in the toughest part of the Creggan estate stemmed very much from the skill and personal example of Major Drummond, who was awarded the MBE.

On 8 August Private Read of 4 Platoon found a 66mm mortar in the boot of a derelict car, which subsequently relieved the Battalion of much harassment from this weapon.

Three stores of bomb-making kit were discovered by alert and inquisitive patrols. One of these finds was largely due to Sergeant Nicholls who noticed a light burning in a house on the Creggan estate which was known to be empty. Nicholls, whose quick wits had led to the capture of a prominent terrorist in Belfast, had been promoted to Sergeant just before the Londonderry tour. A few days later he and Corporal Lonergan arrested a much-wanted terrorist; in the riot that followed, both NCOs succeeded in dispersing the crowd so rapidly that their prisoner could be removed without hindrance. Sergeant Nicholls, who was on his third tour in Northern Ireland, continued to be an inspiration to his platoon and again distinguished himself on 19 November by saving his patrol from taking casualties as a result of his quick reaction and spirited follow-up of a gunman who had opened fire at close range. For his consistently good work over three years, he was awarded the BEM for gallantry.

On 17 September a suitcase full of Provisional intelligence material was found. The girl carrying it escaped, but was caught four days later in possession of parts of an Armalite rifle. She was found to be a hard-core Provisional. By the end of September the Battalion had arrested twelve such terrorists and twenty others, of whom twelve were charged with various criminal offences, plus some fifty rioters.

Against these successes had to be balanced the Battalion's casualties, mainly from stone-throwing rioters. Among them was Second-Lieutenant Michael of 'C' Company, who was hit on the head on 30 July and had to be placed on the Very Seriously Ill list. As the rioting was brought under control, so the shooting increased. In mid-October Sergeant Lenton twice came under automatic fire while leading patrols but with great coolness returned the fire, arranged the evacuation of a wounded soldier, and led his patrol in a quick follow-up of the gunmen. Six soldiers were wounded in October and November: Corporal Seers, 'A' Company, Privates Wells, Stokes, Grivens and Finney of 'C' Company, and Private Sheppard of Support Company. Near misses were also numerous, the most noteworthy being Lance-Corporal Chapman, 'B' Company, who had an Armalite round strike his helmet, passing through one side without touching his head, and Private Brockbank, Support Company, who had his eyebrow nicked by a bullet.

The worst moment of the Battalion's tour came on 3 October, as Colonel Thorne describes:

> 'At 14.30 CLF (Major-General Leng*) and Comd 8 Bde (Brigadier Mostyn) came into the Bligh's Lane base to visit the Essex Factory; knowing of this visit we were on our toes for a rocket attack. At 15.12 there was an explosion in 'C' Coy Ops Room which killed Second-Lieutenant Dobbie (RAOC, attached to 'C' Coy) and very seriously wounded Captain Ray Hazan (2IC 'C' Coy).
>
> 'Support Coy communications also went dead and, thinking there had been a rocket attack, Easy Rider (Emergency road block SOP) was put into effect. On my getting to Bligh's Lane around 15.20 the scene was very confused; however a quick check of the buildings revealed no rocket strike and this fact coincided almost immediately with information which disclosed that a small parcel had been taken to 'C' Coy Ops Room at the time that the CLF was in the Essex Factory—two minutes after handing it in the explosion occurred.
>
> By now the Headmaster of St Joseph's School had let his youths out of school thirty minutes early—rioting was gathering momentum. I called off Easy Rider to cool the situation but by 16.00 the Headmaster had rung the Secretary of State's Office and the BBC and announced the closure of his school because of "Army harassment". The only good thing that day was the picking up of one John "Smokey" Deery, the QM for the Creggan Provos, by a special patrol led by Corporal Beeby (Recce Platoon).'

In the fire that followed the explosion in the Operations Room, Colour-Sergeant Groom and Sergeant Wright ran into the flames and brought out Captain Hazan, possibly

* Formerly Commanding Officer of the 3rd Battalion in Berlin and Aden.

3. When snipers started to operate in Northern Ireland, the umbersome anti-riot shields and batons (*see jacket picture*) were iscarded. Normal patrol tactics for street fighting became the order of ie day.

4. The car-bomb featured prominently in 1973. Innocent car owners ere often forced to carry a time-bomb to a pre-arranged location under ireat of 'execution'.

15. A 1st Battalion soldier on snow shoes clad for cold weath warfare.

16. Snowcat.

7. Ideal country for armoured training: 3rd Battalion soldiers dig in at
uffield, Canada in 1974.

3. Tiger Company marching through Leicester on 25 July, 1974,
xercising the Freedom privilege conferred on The Royal Leicestershire Regiment.
he Colours are those of the 1st Battalion, The Royal Leicestershire Regiment.

19. In the splendid tattoo arena at Tidworth the new Colours are piled ready for The Presentation on 12 July, 1974.

20. At one time the Regiment provided three Commanding Officers of SAF Battalions; (*left to right*) Lt-Cols Turnill, Mackain-Bremner and Worthy find time to enjoy seconded service traditions.

saving his life. Nevertheless, Captain Hazan was totally blinded and severely maimed. Colonel Thorne goes on to describe the sequel to this tragic incident:

> 'I spent the next two days being as reasonable as I could with the Headmaster! In the end he opened his school again on Monday (the 8th). On 4 October an employee at the Essex Factory, Mr James Brown, collapsed and was given artificial respiration by two of our soldiers who saved his life. This had a considerable impact in the surrounding estate when contrasted with the attitude of the workers the day before—women laughing—men playing football close by.'

As October drew to its close the Battalion evened the score somewhat with the identification and arrest by soldiers of 'A' Company of two more hard-core Provisionals, Willie McDaid and Paul Carlin. The last three weeks of the tour saw a series of preplanned searches into various houses in the Creggan, which produced four rifles and 12,000 rounds of ammunition.

One of the characteristics of the Ulster operations has been the presence, never before experienced to the same extent or in such close proximity, of the Press and more notably the television camera. Officers and NCOs have had to become accustomed, without warning and often in the midst of operations, to having a microphone thrust under their noses with the cameras recording their every action. While this instant reporting has done much to bring home to the British public the dangers and difficulties of their soldiers' task in Ulster, it can also be burdensome, time-consuming and at times irresponsible and counter-productive to the Army's efforts. An example of the latter affected the 2nd Battalion, when, in October, 1973, a news team came into the Creggan to film the effect on the children of the estate of the violence and the constant military presence. By this time the Battalion had got the rioting in the Creggan firmly under control; there was thus none for the crews to film. They therefore engineered one, by enlisting the son of a leading Provisional, who soon produced a mob of stone-throwing youths. Colour-Sergeant Underwood's platoon were being stoned heavily when, with matters getting out of control for the first time in weeks, Underwood spotted the camera crews. Lieutenant-Colonel Thorne ordered their arrest which immediately ended the riot—and the filming.

After the tour, Lieutenant-Colonel Thorne received an operational award of the OBE. The citation recorded that, under his command, the Battalion had become 'a highly professional unit with high morale' and went on 'for four months Lieutenant-Colonel Thorne has never spared himself for an instant. By his leadership, drive and indefatigable personal example, he has achieved more than could possibly be expected of him'.

Two other immediate awards of the MBE went to officers of the Regiment in 1973. Major Dodd received his for service while attached to the Headquarters staff at Lisburn; Captain Menage was the Military Intelligence Officer with a RUC Police Division, where, in the words of the citation,

> 'he built up an encyclopaedic knowledge of the enemy in and around the Division. He was tireless in his efforts to ferret out the facts and contact witnesses . . . he made a singular contribution to the operational success of the area.'

So, as November drew to its close, the Battalion's third tour, and the Regiment's sixth, came to an end. For their part, the battalions of the Regiment could look back on a task well and professionally done, a demanding, dangerous and often thankless task, carried out in what one newspaper has described as 'the full glare of public apathy'*. Yet not entirely thankless, as these verses written by a child in a Belfast primary school, touchingly show:

> 'Out in this cold, the wind and the rain
> Taunted and ridiculed time and again

* *The Observer*, 8 December, 1974.

Vainly endeavouring peace to maintain—
Who are you?—a British soldier.

Yours is a debt we can never repay.
Nobody knows how long you must stay,
From morning to evening, from day to day,
God bless you! Oh, British soldier.'

CHAPTER 10

TEN YEARS ON

'In good fortune and in ill it is rare indeed that a British regiment does not hold together; and this indestructible cohesion, best of all the qualities that an armed body can possess, is based not merely on hereditary resolution, but on mutual confidence and mutual respect.'

G. F. R. Henderson, *The Science of War*.

THE PRECEDING CHAPTERS have taken the story, as far as the 2nd, 3rd and 4th Battalions are concerned, up to the end of 1973. Before concluding with the tenth year of the Regiment's existence, it is necessary to go back to 1972 to see how the 1st Battalion fared in Cyprus.

Having left Londonderry on 11 March, 1972, the Battalion was complete in Cyprus by 20 April. Since the 2nd Battalion's days there in 1964–67, the garrison had been reduced; thus the 1st found itself as the only infantry battalion in Near East Land Forces and indeed the only major field force unit on the island. It was therefore responsible for internal security within both of the Sovereign Base Areas, a task which required a split deployment with the Battalion less Support Company in the Episkopi SBA, and the latter Company fifty miles away in Dhekelia, where the 2nd Battalion had been.

The combined internal security tasks in the two SBAs usually absorbed an average of five platoons' worth of soldiers on duty or stand-by at all times, although in the summer of 1973 the duty company had all three platoons constantly deployed, as were two more platoons from other companies.

These commitments imposed considerable limitations on training. However, by exercising one company at a time and making use of facilities in Kenya and in Sharjah in the Persian Gulf, the Battalion managed to make itself proficient in the skills of airportability, limited war and rural counter-insurgency, skills which had been lost during the years of mechanization and urban guerrilla warfare. Time was also found for adventure training and while in East Africa, a party under Captain Thomas reached the summit of Mount Kenya (17,058 feet).

The experience gained in Ulster paid dividends when it came to competition shooting. In the 1972 NEARELF Rifle Meeting, 1st Battalion teams and individuals completely swept the board, an achievement typified by the performance of Private Sluggett, whose photograph in *Castle* showed him amid the nine silver trophies he personally won at this Rifle Meeting. In the 1973 Meeting, the three companies, competing as minor units, won every team event, except the falling plate and the pistol, against such stern opposition as 41 Royal Marine Commando and the Canadian United Nations Battalion. Lance-Corporal Cheong was the decisive NEARELF Rifle Champion. Fired with this success, the Battalion sent a team to Bisley, where the falling plate team of Corporal Denny, Lance-Corporals Cheong and Wharton and Private Sluggett made up for their failure in Cyprus by winning the Royal Ulster Rifles Cup in a match with 143 entries. Cheong won the Army Hundred Cup and was joined in the Hundred by Sluggett, with Wharton and Major Long just failing to qualify with scores of 106 and 112 respectively.

The Battalion had been taken to Cyprus by Lieutenant-Colonel Jackson but shortly afterwards, on 10 June, 1972, he handed over to Lieutenant-Colonel D. C. Thorne, the twin brother of M. E. Thorne, who assumed command of the 2nd Battalion in March, 1973. To have twins commanding two battalions of the same regiment at the same time must be fairly rare in the history of the Army.

The Battalion Band made good use of Cyprus as a stepping-off ground with visits in 1972 and 1973 to Turkey and Israel, where the 2nd Battalion Band had enjoyed a successful tour in 1966. A proposed visit to the Lebanon in 1973 had to be cancelled, but a musical safari to Kenya made an adequate substitute. This culminated in a Massed Bands performance at the Nairobi Agricultural Show in which the Battalion Band joined those of the Kenya Army, Police and Prison Service, a total of 600 musicians.

The Battalion's sojourn in Cyprus lasted only two years and in May, 1974, an entirely new rôle beckoned—that of the British battalion in the Land Component of the Allied Command Europe Mobile Force (AMF(L)). This force owed its origin to the NATO policy of 'flexible response', which ensured a graduated reaction by the Western Alliance to the threat of a Soviet confrontation. Certain areas of Western Europe, the northern and southern flanks of the Alliance and Northern Italy are particularly vulnerable, owing to their lack of NATO forces or the existence of a common frontier with Russia or her satellites. To counter the danger to these areas within the 'flexible response' policy, a multi-national force was set up in March, 1960. This could be moved rapidly to any threatened area and would, by its very composition and presence, aim to deter an aggressor on the NATO principle of an attack against one member being an attack against all. The contributing nations to this force are Belgium, Canada, Germany, Italy, Luxembourg, the United States and the United Kingdom; the latter provides an infantry battalion, an artillery battery, a radio troop, a reconnaissance squadron, a helicopter unit, a force artillery headquarters, and certain elements of the Logistic Support Battalion, together with a number of staff officers at the Force's Headquarters in Germany. When the 1st Battalion joined the Force, the Regiment already had two representatives at the headquarters—Colonel Turnill, the Chief of Staff, and Lieutenant-Colonel Gill, G2 (Chief Intelligence Officer).

The 1st Battalion's base for this undertaking was Tidworth, where it arrived in May, 1974, with the prospect of a new and exciting rôle before it. But first there was a different enterprise to prepare for, one requiring skills of historic precedent, as will transpire later.

After its highly successful third tour in Ulster, the 2nd Battalion returned to Munster to pick up once again the threads of mechanized soldiering. For the first time since going to BAOR the Battalion could look forward to a complete year of throwing the battle groups around the North German plain, uninterrupted by the demands of urban guerrilla warfare. Once more the specialist cadres were set in train, followed by the time-honoured progression of individual, section, platoon, combat team, battle group and finally brigade training in the autumn. So the 2nd Battalion again settled into the familiar BAOR routine, with its year punctuated, not so much by the months or seasons, but by the events cryptically described by those groups of initials so dear to the military mind of the seventies—CPX, NCE, FTX, CIV, FFR and so on.

Freedom from the Ulster commitment, or 'Op Banner' as it was known, enabled the Battalion to field a team for the Major Units Championship at Bisley for the first time. In the snapshooting match of the Championship the team gained 5th place out of the seventy units competing, but overall the lack of previous Bisley experience told against them, and the final result of the Championship put the Battalion in 40th position. Competing against them was the 1st Battalion team, which managed to better its result in the 1973 meeting by moving up from 15th to 10th. Lance-Corporal Sluggett again achieved the Army 100, being joined by Corporal Denny and Lieutenant Domeisen. The latter had a triumphant shooting year, winning the Roupell Cup, the Old Contemptibles Cup, and the Queen Mary Cup, shooting for the Army VIII, and also being champion of South-Western District and Northern Ireland.

Life in 1974 for the 3rd Battalion in Paderborn began with

a change of commanding officers, Lieutenant-Colonel Hall-Tipping handing over to Lieutenant-Colonel Aris. In May the prospect of a further tour in Northern Ireland suddenly loomed, then died away leaving the Battalion free to continue the BAOR cycle. In the summer two companies had a change from the familiar terrain of Soltau and Sennelager. 'C' Company enjoyed a month in the French training area of La Courtine, while 'A' Company, with Battalion Headquarters, two squadrons of the 9th/12th Lancers, a battery of 2nd Field Regiment RA, and 'S' Troop of 2nd Armoured Engineers Squadron, formed a battle group exercising in the 1,000-square mile field firing area of Suffield in Canada. Suffield had been established as a training area in 1972 to provide realistic training for battle groups from the United Kingdom and Germany, where the areas are too confined to permit simultaneous live firing of all weapons in a battle group. The experience was greatly appreciated by the 3rd Battalion men who took part and afforded not only such unrestricted training as few had ever undergone before, but also an opportunity to visit other parts of Canada and the United States.

Tiger Company were last met at the end of Chapter 6 reforming as an operational company at Canterbury. In March, 1974, after the customary preliminary training, the Company moved to Northern Ireland, thus completing the appearance on that stage of the entire Regular element of the Regiment. After a quiet two months looking after Portadown and the surrounding countryside, the Company, less one platoon detached at Dungannon, moved to Armagh and came under command of the Life Guards, with the task of patrolling the border between the province and the Republic. The worsening political situation caused by the Ulster Workers Council strike saw the Company withdrawn to brigade reserve, where it spent the remainder of the tour until returning to Canterbury in early July. Except for the platoon at Dungannon, which had a more active rôle, the Tigers had had a fairly uneventful stay and, apart from one night's operation in Lurgan, had never been allowed to function as a complete company, which was disappointing.

On 25 July the Company was blessed with fine weather and a large audience to witness its march through Leicester to exercise its right to the Freedom of the City. Then, after a change of company commanders, Major Growse taking over from Major Swallow, it was back to the first rôle the Company had undertaken as an independent body, that of demonstration troops at the Royal Military Academy, Sandhurst.

Throughout 1974 the Volunteer battalions increased in strength and skill, although the need to attract more recruits remained constant and the efforts to do so unremitting. Lieutenant-Colonels Raywood and Wallace, who had launched the 6th and 7th Battalions in 1971, handed over to Lieutenant-Colonels Swayne and Dawson respectively. Lieutenant-Colonel Swayne, a skilled marksman, lent his expertise to the 6th Battalion's shooting team, which, under the experienced tuition of Regimental Sergeant-Major Knight, made its first appearance at Bisley, achieving 3rd place in the China Cup, the championship match for TAVR units. In the individual TAVR competition, Private Marshall came equal 5th and Lieutenant-Colonel Swayne equal 9th out of a total of about 700. These results reflected great credit on the Permanent Staff Instructors, led by Major Wilson, and presaged further successes for the 6th Battalion at Bisleys yet to come. As the year proceeded all three Volunteer battalions developed a high pitch of military and social activity, building up to their Annual Camps, held, in 1974, in the United Kingdom; a member of the 5th Battalion wrote: 'As always life has been hectic'.

The life of a modern regiment revolves, of necessity, around modern matters: around APCs, urban guerrillas, Fitness for Rôle Inspections, sophisticated weaponry, career planning, the myriad complexities of administration and logistics. There is constant preoccupation with the problems of today, tomorrow and the future, whereas in days gone by the daily life of a regiment varied little in essence from decade to decade, even from century to century, giving many British regiments a deeply conservative attitude to life, steeped in precedent and the past. A regiment's good name often stemmed from some glorious action on some bloody field a century before; nowadays its prowess is judged on its current efficiency and the fact that it once saved or lost its colours in some forgotten skirmish in a border defile—circumstances that could make or mar a regiment's reputation for generations—is of little account. Modern regiments are concerned with modernity.

Yet even the most up-to-date, the most forward-looking of regiments can still find time for past traditions and undertakings of ancient origin. In the Royal Anglian Regiment the hours of past glory are still celebrated annually on the anniversaries commemorating its chief honours. In 1974 parties from the 3rd Battalion visited two fields where their ancestors, the 16th and 44th Foot, achieved fame—the fields of Blenheim and Salamanca. In 1974 too, the tenth year of the Regiment's existence, there was time for a ceremony whose origins pre-dated Blenheim. On 12 July, the Colonel-in-Chief of the Regiment, Her Majesty Queen Elizabeth The Queen Mother presented Colours to four Battalions of the Regiment.

On formation in 1964 the Regular battalions had inherited the Colours presented to the 1st, 2nd, and 3rd East Anglian Regiments in 1963, 1962 and 1959 respectively; the 4th Battalion had continued to carry the Colours presented to 1st Battalion The Royal Leicestershire Regiment in 1954 and which, in 1974, were in the keeping of Tiger Company. The 5th Battalion had not been presented with Colours and the policy on Colours for the 6th and 7th Battalions had not been decided. Now the Royal Anglian Regiment was to receive its own Colours for the first time.

With the 2nd and 3rd Battalions in Germany, it fell to the 1st Battalion, stationed at Tidworth, to make the arrangements for this important occasion; in particular upon the

Commanding Officer, Lieutenant-Colonel D. C. Thorne, the Adjutant, Captain Walker, the Regimental Sergeant-Major, WO 1 Buffine, and upon Major Growse, whose vital task lay in the planning and execution of the highly detailed administrative preparations required to make this royal occasion the impressive, smoothly-functioning affair it turned out to be. The Parade was to be formed of four guards provided by the 1st Battalion under the command of Lieutenant-Colonel Thorne, the Massed Bands and Drums of the three Regular battalions under the senior Drum-Major, WO 2 West of the 2nd Battalion, and Colour Parties of the Regular battalions and the 5th Battalion;* the Commanding Officers, one Major, and the Quartermaster of the 2nd, 3rd and 5th Battalions were also in attendance. Large numbers of guests of every degree were bidden to attend, dignitaries of town and county, senior officers, representatives of Allied regiments, including the Commanding Officers of the Lake Superior Scottish, the Bermuda and Gibraltar Regiments, who came all the way from their respective countries, and past members of the Regiment and former Regiments. The Colonel of the Regiment, Lieutenant-General Sir Ian Freeland, and the Deputy Colonels assembled; and a body of clergy, headed by the Chaplain-General to the Forces, Archdeacon Mallet, stood ready to consecrate the Colours. So, on to the greensward of the Tidworth polo ground within its ring of trees, to the heart-lifting strains of 'Sambre et Meuse', marched the Parade in khaki No. 2 Dress, with Bands in No. 1 Dress and Drums resplendent in scarlet, to receive its Colours from its Colonel-in-Chief.

Her Majesty arrived in modern style by helicopter of the Queen's Flight and inspected the Parade. The troops then formed a hollow square, the drums were piled and the four

* Colour Parties. 1st Bn: Lieuts Mooring and Stallard, WO 2 Beckett, C/Sgts Carpenter and Pryce; 2nd Bn: Lieuts Deed and Hodgson, WO 2 Keal, C/Sgts Underwood and Whitehead; 3rd Bn: Lieuts Field and Duthoit, WO 2 Graham, C/Sgts Lowe and Bryant; 5th (V) Bn: Capt George, Lieut Carr, WO 2 Burgess, C/Sgts Lee and Lacey.

sets of new Colours were marched into position for the service. With the Colours duly consecrated, the Queen Mother presented each set in turn to the Colour parties and then addressed the Parade. The Colonel of the Regiment replied, the four guards re-formed their line, and the Colours were marched to their positions in the line. The Parade next marched past the saluting base in slow and quick time. As the Colours went by the Colonel-in-Chief, the ensigns let them fly, displaying the richness of the coloured silk, emblazoned with the serried battle honours—Ramillies, Quebec, Waterloo, Inkerman, Afghanistan, Ypres, Imphal, Korea—41 on each Queen's Colour, 45 on the Regimental, distinctions gained over three centuries by generations of soldiers past. It is a pleasing fancy to believe that the ghosts of these old warriors observed with critical though indulgent eyes the marching column of their descendants. None knew better than they the truth of the words of a distinguished officer of the old Army and still applicable, perhaps, to the modern infantryman:

> 'One way or another the roots of tradition strike down deep. The soldier feels the Regiment solid about him.'*

The march past completed, the Parade advanced in review order to the 'British Grenadiers' and presented arms. The National Anthem was played, and then the caps came off and the Royal Anglians gave three cheers for their Colonel-in-Chief. Finally, as countless old soldiers of the 9th and 12th Foot had done before, the Parade marched off to 'Rule, Britannia' and 'Speed the Plough'. It was truly a magnificent occasion, a splendid modern rendering of an historic ceremonial, worthy of the Regiment's forbears and a fitting celebration of the Regiment's first decade.

So, as the tenth anniversary of the formation approached, and with it the close of this story, the battalions of the Royal Anglian Regiment, efficient, well-trained, seasoned infantry,

* General Sir Ian Hamilton, *The Soul and Body of an Army* (1921).

stood ready for whatever the next decade might hold. It will be recalled that in 1962, when the concept of the 'Large Regiment' was first announced, it found little favour with many regiments of infantry. So the tenth anniversary of the first of the Large Regiments is perhaps a suitable time to assess the success of the venture.

Royal Anglian battalions have performed with credit and often distinction in the two theatres of active operations the Army has had to contend with during the decade—Aden and Northern Ireland. The Regular battalions have all proved themselves as mechanized infantry, the skills for which, in recent years, have been more difficult to maintain owing to the simultaneous need to acquire the different skills of urban guerrilla warfare. The battalions have earned gratifying reports for their standard of training and administration from inspecting officers and the loss of county identities does not seem to have hindered recruiting. But, it might be argued, other regiments which withstood the 1962 proposals have performed equally creditably in these spheres, so the test of the Royal Anglians' decision to accede to those proposals must lie elsewhere.

The story, as it has unrolled, seems to prove that the aim of the concept given by the Bower Report has been fulfilled: the essential features of the regimental system, its traditions and territorial links have been maintained; the Regiment has lost a Regular battalion and gained two Volunteer battalions without undue stress. The problems of reinforcing battalions and planning the careers of officers and NCOs have been greatly simplified by the common title and common rolls; drafting thirty men from the 3rd Battalion to the 1st has caused far less heart-searching and anxieties than would have been entailed in drafting from one single-battalion regiment to another. The institution of the Divisional system has inflicted many of the characteristics of the Large Regiment on all regiments, but in the Queen's Division, with three regiments each of three battalions, a soldier faces much less chance of changing his badge and loyalties than in, for example, the Prince of Wales's Division with nine single-battalion regiments. In any case, when the decision to form the Regiment was made, the Divisional system had not been devised so its introduction is no argument against the Royal Anglian's decision.

The real test of success or failure of the Regiment's formation seems to be one of identity and reputation—whether it has succeeded in establishing itself as an integrated whole, in which its members can take pride and feel they are part of a regimental family, and which the Army and the public fully acknowledge, with a comparable status to that of the old county regiments. Of course the change to a larger grouping has involved some loss of identity and of old traditions and customs. But such things mean more to the old soldier than his serving counterpart, whose loyalties are more concerned with the present and the efficient functioning of the battalion in which he finds himself; indeed, the comparative 'newness' of the Regiment and its competence in the Bogside is almost certainly a greater attraction to a young man, officer or soldier, than the fame of its ancestors at Malplaquet or Talavera. A senior officer, asked for an objective view of the Royal Anglian enterprise, has said:

> 'In my opinion it has been an unqualified success in having created a genuine entity. There are temporary loyalties to one battalion or another, especially among those who have served with one more than another. This is natural, indeed desirable, but I have heard no one complain at postings to a particular battalion because of its number or reputation. People do feel they are Royal Anglians and I think the Regiment has earned some envy for this from less cohesive attempts at "Large Regimentation".'

By its conduct in Ulster and the training grounds of the Army's world, its successes at sport and shooting, its officers who have attained senior rank in the Army* (and perhaps by

* Of those officers who, on the formation of the Regiment in 1964 were of the rank of Lieutenant-Colonel and below, by 1 September, 1974, three were Major-Generals, three were Brigadiers and ten were Colonels. One Lieutenant-Colonel was promoted to Brigadier shortly afterwards.

its very size and the passage of time), the Regiment has achieved full recognition of its identity in the Army as a whole; the days when a Royal Anglian was automatically questioned as to his former regiment are past. In the counties from which its recruits are drawn, its status as the successor to the former regiments of those counties is now fully appreciated and understood. As for the general public, while the title of the Regiment may not elicit such instant response as is given to, say, the Grenadier Guards or the Black Watch, its name is now at least as familiar as that of many older Regiments of the Line, which, from a public which has never taken great interest in its Army except in war, is as much as can be expected.

Throughout the 300-odd years of Britain's standing Army, the brunt of the nation's military operations has been borne by the Regiments of the Line. Seldom accorded its due esteem, lacking perhaps the glamour of the Household troops or the cavalry, the British infantry has nevertheless earned an enviable reputation for stolid endurance, dependability and courage. By striving to live up to this reputation the Royal Anglian Regiment has earned its place and maintained the traditions of its forbears.

APPENDIX I

BATTLE HONOURS

The Regiment inherited from the former regiments a total of 299 Battle Honours: 69 for the period up to 1914; 102 for the Great War; 126 for the Second World War; and two for the Korean War. Many were common to more than one of the former regiments. Of these the following were selected for emblazoning on the Colours:

War in the Spanish Netherlands 1689–97
NAMUR 1695

War of the Spanish Succession 1702–12
BLENHEIM, RAMILLIES, OUDENARDE, MALPLAQUET

War of the Austrian Succession 1713–47
DETTINGEN

Seven Years War 1756–63
LOUISBURG, MINDEN, QUEBEC 1759, MARTINIQUE 1762, HAVANNAH

French Revolutionary, Napoleonic, Indian and American Wars 1793–1815
MARTINIQUE 1794, SERINGAPATAM, CORUNNA, TALAVERA, ALBUHERA, BADAJOZ, SALAMANCA, VITTORIA, PENINSULA, BLADENSBURG, WATERLOO

Colonial Wars 1816–54
AVA (1824), GHUZNEE 1839, KHELAT (1839), CABOOL 1842, MOODKEE (1845), FEROZESHAH (1845), SOBRAON (1846), NEW ZEALAND (1846), GOOJERAT (1849), PUNJAUB (1849), SOUTH AFRICA 1851–53

Crimean War 1854–56
INKERMAN, SEVASTOPOL

Colonial Wars 1856–98
LUCKNOW (1857), TAKU FORTS (1860), AFGHANISTAN 1878–80,

KABUL 1879, SOUTH AFRICA 1879, NILE 1884–85, TIRAH (1897), ATBARA (1898), KHARTOUM (1898)

South African War 1899–1902

PAARDEBERG, DEFENCE OF LADYSMITH, SOUTH AFRICA 1899–1902

Great War 1914–18

MONS, LE CATEAU, MARNE 1914, AISNE 1914 1918, YPRES 1914 1915 1917 1918, NEUVE CHAPELLE, LOOS, SOMME 1916 1918, ARRAS 1917 1918, CAMBRAI 1917 1918, FRANCE AND FLANDERS 1914–18, MACEDONIA 1915–18, GALLIPOLI 1915–16, GAZA, PALESTINE 1917–1918, SHAIBA, MESOPOTAMIA 1914–18

Second World War 1939–45

ST OMER-LA BASSEE, DUNKIRK 1940, NORMANDY LANDING, BRIEUX BRIDGEHEAD, VENRAIJ, N.W. EUROPE 1940 1944–45, TOBRUK 1941, DEFENCE OF ALAMEIN LINE, NORTH AFRICA 1940–43, VILLA GRANDE, SALERNO, ANZIO, CASSINO I–II, GOTHIC LINE, ITALY 1943–45, CRETE, SINGAPORE ISLAND, MALAYA 1941–42, YU, NGAKYEDAUK PASS, IMPHAL, KOHIMA, CHINDITS 1944, BURMA 1943–45.

Korean War 1950–53

KOREA 1951–52

Notes:

1. Dates in brackets are not included on the Colours.
2. Martinique 1794 is included with Martinique 1762 on the Regimental Colour.
   Kabul 1879 is included with Cabool 1842 on the Regimental Colour.
   South Africa 1879 is included with South Africa 1851–53 on the Regimental Colour.

APPENDIX II

LIST OF COLONELS OF THE REGIMENT, DEPUTY COLONELS AND COMMANDING OFFICERS

COLONELS OF THE REGIMENT

1964–1966	Lieutenant-General Sir Reginald Denning KBE CB DL
1966–1971	Lieutenant-General Sir Richard Goodwin KCB CBE DSO DL
1971–	Lieutenant-General Sir Ian Freeland GBE KCB DSO DL

DEPUTY COLONELS OF THE REGIMENT

1964–1971	Lieutenant-General Sir Richard Goodwin KCB CBE DSO
1964–1965	Major-General Sir Douglas Kendrew KCMG CB CBE DSO
1964–1966	Brigadier R. H. L. Oulton CBE
1964–1970	Brigadier C. M. Paton CVO CBE DL
1965–1971	Major-General J. M. K. Spurling CB CBE DSO
1966–	Brigadier P. W. P. Green CBE DSO
1966–1971	Major-General I. H. Freeland CB DSO
1970–	Major-General M. W. Holme CBE MC
1971–	Colonel M. St G. Pallot
1971–1974	Major-General G. R. Turner-Cain CB CBE DSO
1974–	Major-General J. B. Dye CBE MC

REGIMENTAL COLONELS

1964–1966	Colonel J. C. Denny OBE MC
1966–1968	Colonel P. E. B. Badger

(*Formation of The Queen's Division*)

COMMANDING OFFICERS

1st Battalion

1964–1965 Lt-Col J. B. Dye OBE MC
1965–1967 Lt-Col T. M. Creasey OBE
1967–1969 Lt-Col B. H. C. Emsden
1969–1972 Lt-Col R. L. Jackson OBE
1972– Lt-Col D. C. Thorne

2nd Battalion

1964–1965 Lt-Col W. R. Chambers
1965–1968 Lt-Col I. A. Haycraft
1968–1970 Lt-Col J. B. Akehurst
1970–1973 Lt-Col R. E. J. Gerrard-Wright OBE
1973– Lt-Col M. E. Thorne OBE

7th (Volunteer) Battalion

1970–1974 Lt-Col W. G. Wallace OBE TD

The Depot

1964–1967 Lt-Col H. W. Clark
1967–1969 Lt-Col J. V. Miseroy
1969– Lt-Col W. C. Deller OBE
(*Formation Queen's Division Depot*)

3rd Battalion

1964– Lt-Col C. C. Norbury MC
1964–1966 Lt-Col P. J. H. Leng MBE MC
1966–1969 Lt-Col J. L. M. Dymoke MBE
1969–1971 Lt-Col K. Burch MBE
1971–1974 Lt-Col J. Hall-Tipping
1974– Lt-Col M. A. Aris

4th Battalion

1964–1966 Lt-Col J. A. C. Cowan MBE
1966–1968 Lt-Col D. R. C. Carter
1968–1970 Lt-Col T. Holloway
(*Reduced to Company strength*)

5th (Volunteer) Battalion

1965–1968 Lt-Col H. H. Moore
1968–1971 Lt-Col P. D. Blyth
1971–1973 Lt-Col J. R. Heath
1973– Lt-Col P. D. L. Hopper

6th (Volunteer) Battalion

1970–1973 Lt-Col P. W. Raywood TD
1973– Lt-Col T. C. B. Swayne TD

APPENDIX III

HONOURS AND AWARDS

This list of honours and awards has been compiled from information available in Regimental Records, Army Lists and the *London Gazette*. It is realized that the list may not be complete and it is hoped that every reader of the Appendix whose name has been inadvertently omitted or who has any other correction to make will inform The Regimental Secretary, Gibraltar Barracks, Bury St Edmunds, Suffolk.

Awards are listed in chronological order and where they were made for gallantry or meritorious service they are marked:

† for Gallantry * for Meritorious Service

Knight Grand Cross of The Most Excellent Order of The British Empire (GBE)

Lieutenant-General Sir Ian Freeland KCB DSO

Knight Commander of The Most Honourable Order of The Bath (KCB)

Lieutenant-General I. H. Freeland CB DSO

Lieutenant-General M. A. H. Butler CB CBE DSO MC

Companion of The Most Honourable Order of the Bath (CB)

Major-General G. R. Turner-Cain CBE DSO

Commander of The Most Excellent Order of The British Empire (CBE)

Brigadier J. B. Dye OBE MC

Colonel R. P. S. Erskine-Tulloch

Colonel D. S. Carden

Brigadier M. W. Holme OBE MC

Officer of The Most Excellent Order of The British Empire (OBE)

Lieutenant-Colonel J. B. Dye MC
Lieutenant-Colonel D. U. Fraser
Lieutenant-Colonel T. M. Creasey
Lieutenant-Colonel J. S. Greene MC
*Lieutenant-Colonel J. M. Petit MBE
Lieutenant-Colonel R. E. J. Gerrard-Wright MBE
Lieutenent-Colonel E. Turnill
Lieutenant-Colonel W. C. Deller
Colonel G. C. Howgego
Lieutenant-Colonel A. F. Mackain-Bremner
*Lieutenant-Colonel R. L. Jackson
*Lieutenant-Colonel M. E. Thorne

Member of The Most Excellent Order of The British Empire (MBE)

†Lieutenant W. J. Hawkins
Major K. Burch
Captain N. R. Pavitt
Major R. E. J. Gerrard-Wright
WO2 T. W. Ward BEM
WO2 D. G. Groves
Major W. D. G. Fairholme
Captain (QM) A. D. Russell
†Lieutenant B. A. R. Harrington-Spier
Lieutenant N. H. P. Jenks MM
Major (QM) D. J. Bebbington
Captain (QM) R. Bayliss
WO1 (Bandmaster) B. Harper
†Captain J. R. Hart
Major R. H. Robinson
Major K. J. T. Hoile
*Captain A. E. Thompson MC
†Major C. M. J. Barnes
†Captain R. Howe
*Major G. H. Bradshaw
*Major W. T. Dodd
*Major R. J. M. Drummond
*Captain M. J. Menage
Lieutenant-Colonel P. J. Sincock
WO1 G. E. Veitch

†*The Military Cross (MC)*

Captain S. M. Brogan

†*The Distinguished Conduct Medal (DCM)*

Sergeant N. Whitfield

†*The Military Medal (MM)*

Corporal N. Jephcote
Staff-Sergeant A. J. Underwood
Lance-Corporal W. J. Simpson

The British Empire Medal (BEM)

Sergeant J. Watson
†Sergeant J. W. Jones
†Private J. V. Elba-Porter
†Private A. D. Kent
†Sergeant W. Allan
†Corporal J. M. Valentine
†Staff-Sergeant M. E. Nugent
†Sergeant R. W. Nicholls

The Meritorious Service Medal (MSM)

WO1 (RSM) B. Driscoll
WO1 (ORQMS) E. G. A. Smith

Mentioned in Despatches

Lieutenant P. B. D. Long
Lance-Corporal P. D. R. Watson
Lieutenant C. J. Brock
Staff-Sergeant B. Hutchinson
Corporal P. V. Townsend
Sergeant P. L. Lane
Corporal P. Tatlock
Colour-Sergeant D. S. Wade
Colour-Sergeant R. Smith
Sergeant D. J. Foran
Corporal P. L. Ludbrooke
Lieutenant-Colonel J. Hall-Tipping (2)
Major T. D. A. Veitch
Major P. P. Young
Major N. J. Lewis
Lieutenant-Colonel R. E. J. Gerrard-Wright OBE
Lieutenant T. Longland
2nd Lieutenant M. G. A. Shipley
Lieutenant A. Behagg
Sergeant G. W. Wright
Captain A. E. Thompson MBE MC

Queen's Commendation for Brave Conduct

Private H. West
Private A. W. Stamps
Corporal R. C. Day
2nd Lieutenant B. W. Copping
WO2 W. E. Holden
Private M. Richardson
Staff-Sergeant T. A. Barker

AWARDED FOR SERVICE ON SECONDMENT

OMAN

The Sultan's Order of Oman

Major-General T. M. Creasey OBE

The Sultan's Bravery Medal

Lieutenant-Colonel E. Turnill

The Sultan's Distinguished Service Medal

Lieutenant-Colonel A. F. Mackain-Bremner
Lieutenant-Colonel P. Worthy
Major P. J. Sincock

The Sultan's Commendation

Captain P. R. D. Aplin

JORDAN

The Hashemite Order of Independence

Major-General T. M. Creasey OBE

AWARDED TO MEMBERS OF THE TERRITORIAL AND ARMY VOLUNTEER RESERVE

Knight Commander of The Most Excellent Order of the British Empire (KBE)

Colonel A. Noble CB DSO TD DL

Officer of the Most Excellent Order of the British Empire (OBE)

Lieutenant-Colonel F. G. Barber TD
Lieutenant-Colonel R. G. Wilkes TD
Major S. C. Aston TD
Lieutenant-Colonel W. G. Wallace TD

Member of the Most Excellent Order of the British Empire (MBE)

Major D. C. Purves TD

APPENDIX IV

ALLIED REGIMENTS

CANADA

The Lake Superior Scottish Regiment
The Sherbrooke Hussars
The Lincoln and Welland Regiment
The Essex and Kent Scottish

AUSTRALIA

The Royal Tasmania Regiment

NEW ZEALAND

3rd Bn [Auckland (Countess of Ranfurly's Own) and Northland]
Royal New Zealand Infantry Regiment

PAKISTAN

5th Bn The Frontier Force Regiment

MALAYSIA

1st Bn The Royal Malay Regiment

COMMONWEALTH FORCES

The Barbados Regiment
The Bermuda Regiment
The Gibraltar Regiment

APPENDIX V

THE REGIMENT AND THE SULTAN'S ARMED FORCES IN MUSCAT AND OMAN

The dawn of 1 September, 1964, was similar to any other as far as those East Anglians who were serving in the Persian Gulf at this time were concerned. They could not realize that many members of their new Regiment, the Royal Anglian, would contribute so much to the stability of the area in the coming ten years. At this time Brigadier Holme, later to be Deputy Colonel of the Regiment, was responsible for the security of the Gulf as Commander Land Forces Persian Gulf, based in Bahrain. Further away at Sharjah was the Headquarters of the Trucial Oman Scouts where the DAA and QMG was Major Worthy, and commanding the training depot was Major Dinnin, later to be a security officer with an oil company in Oman.

In Muscat and Oman a rebellion had just been started by a small group against the rule of Sultan Said bin Taimur. Soon the uprising was to be developed by subversive groups in Bahrain and Aden into a war sustained by professional revolutionaries based outside Oman.

The Sultan's Armed Forces (SAF) were to be fully involved in this war for many years to come. One of the first to be launched against the initial rebellion was the Northern Frontier Regiment until recently commanded by Lieutenant-Colonel Daglish whose reconnaissance platoon commander was Captain Swallow; both at the time were still Royal Leicesters. Also serving in the country were Major Wooddisse and Captain Peat, the latter on the strength of the Trucial Oman Scouts at Sharjah.

More Royal Anglians were soon to join on loan the slowly expanding Sultanate forces. Major Raven served there continuously for six years and enjoyed many amusing experiences, including his successful leadership of a small expedition of soldiers with the Sultan's permission, to capture a live white Arabian oryx—the mythical 'Unicorn'. Major Aris and Captain Baily served in the Headquarters at Muscat, while Captain Peele was serving in the Muscat

Regiment, where he was known as 'Captain Shadeed' because of his physical toughness. In October, 1968 the Commander of SAF wrote to the Colonel of the Regiment a letter which was full of praise for these four officers and said: 'I feel I should tell you how pleased I am with the officers of your Regiment'.

In 1969 Lieutenant-Colonel Turnill took over command of the Desert Regiment and in February, 1970 moved to Dhofar where he was outnumbered by the enemy on the jebel and was restricted almost entirely to operations in defence of Salalah on the plain. Soon the position was to change, for on 23 July, 1970 the Sultan's son, Qaboos, replaced his father in a bloodless coup and initiated a programme, not only to increase the size and equipment of his forces, but also to reform and develop the country.

The new Sultan almost immediately gave a fresh impetus to the campaign. By September the first surrenders came in and during October information was received that a complete platoon would give themselves up. A total of nineteen arrived and Colonel Turnill noted that they were well turned out in jungle-green uniforms, Chinese caps and web equipment; he also observed that when he spoke to them they were as restless and tense as a row of tethered falcons, but hard men nevertheless. This group was later to be the nucleus of the forces known as Firqats who were formed to fight for the Government against their former comrades.

In the north of Oman the Muscat Regiment was then commanded by Lieutenant-Colonel Mackain-Bremner, who had the unusual experience of co-ordinating a complete change of currency in the absence of any civil government in his area of responsibility. Apparently many produced vast sums of money, especially Maria Theresa dollars, which had been hoarded for many years and, although used as currency, had not been recognized as such by the Government.

Just before the coup the old Sultan had agreed to the formation of a fourth infantry battalion. Lieutenant-Colonel Worthy, who had returned to Oman, was selected to raise, train and take to war this new battalion, later to be named the Jebel Regiment by Sultan Qaboos. Before assuming his new command he had been the first GSO 1 Plans at HQ SAF, a temporary appointment involving the detailed planning for the build-up of SAF in Dhofar and the subsequent offensive operations.

Part of the Royal Anglian contribution to SAF was now three

of the four commanding officers, and this at one of the most critical times in the war as the first major offensive against the enemy was about to be launched. Help was provided by the British Army in the form of training teams. Captain Brogan served in one of these teams and did much in the eastern area of Dhofar.

At the end of 1970 Lieutenant-Colonel Worthy was sent as the Sultan's Liaison Officer to the Joint Forces Headquarters at Sharjah. There, the lavish organization was described by one cynic as similar to Southwick House prior to D-Day, especially as the Commander had similar weather problems! One of the Trucial Oman Scouts squadrons was commanded by Major Barnett, and in a SAF company was Captain Aplin, who had great difficulty in persuading the Royal Navy that his dhow was legitimate and that he and his crew were not a boatload of inept immigrants. Later, on his return to northern Oman, he was involved in helping the local people on the Jebel Akhdar, the scene of much fighting in 1958–59. On sending a signal to regimental headquarters for a flying doctor to visit the villages to circumcise many of the boys at the inhabitants' request, he received a reply: 'Reference your request for circumcisions—delighted—do any British officers require, query.'

As the size of the Force increased, so did the Headquarters. In 1972 the Dhofar Headquarters was renamed Dhofar Brigade. Major Sincock was the first Brigade Major, and in August, 1974, Brigadier Akehurst assumed command of the Brigade. When the Brigade was formed, the rank of the Force Commander was raised and Major-General Creasey was appointed to this task.

At the close of this history of the first ten years of the Royal Anglian Regiment the Deputy Commander of the Union Defence Force (formerly the Trucial Oman Scouts) was Lieutenant-Colonel Robinson, and in SAF the Regiment provided the General, the Brigadier, Lieutenant-Colonel Dale commanding the Desert Regiment, Lieutenant-Colonel Green the GSO.1, Captain Harrington-Spier on the staff, and Major Shervington serving in the Jebel Regiment. Captain Shipley, who was tragically killed in a helicopter crash in 1975, was also serving in the Jebel Regiment at the time.

There is a long tradition of service by British officers on loan to locally-enlisted troops, dating back to the formation, in the seventeenth century, of the Honourable East India Company's forces, later to become the Imperial Indian Army. The services of Royal

Anglian officers to the Arab forces in the Gulf are part of this tradition, and it is to be hoped that, with positions of responsibility being increasingly assumed by the indigenous population, their efforts to train and lead will not have been wasted.

INDEX